GW01218265

Ab Initio Valence Calculations in Chemistry

AB INITIO VALENCE CALCULATIONS IN CHEMISTRY

D. B. COOK

University of Sheffield

BUTTERWORTHS

THE BUTTERWORTH GROUP

ENGLAND
Butterworth & Co (Publishers) Ltd
London: 88 Kingsway, WC2B 6AB

AUSTRALIA
Butterworths Pty Ltd
Sydney: 586 Pacific Highway, NSW 2067
Melbourne: 343 Little Collins Street, 3000
Brisbane: 240 Queen Street, 4000

CANADA
Butterworth & Co (Canada) Ltd
Toronto: 14 Curity Avenue, 374

NEW ZEALAND
Butterworths of New Zealand Ltd
Wellington: 26-28 Waring Taylor Street, 1

SOUTH AFRICA
Butterworth & Co (South Africa) (Pty) Ltd
Durban: 152-154 Gale Street

First published in 1974

©
D.B. Cook
1974

ISBN 0 408 70551 5

Printed in England by
Fletcher & Son Ltd., Norwich

PREFACE

The calculation of molecular wave functions using an orbital model is changing from being an area of quantum chemistry research in its own right to being a routine "service". An increasing number of non-specialists wish to make use of the results of theoretical calculations on molecular systems and there are a few large program packages in existence for the routine calculation of molecular orbital wave functions. However, unlike NMR or mass spectroscopy which have undergone a similar transition from research to facility, molecular calculations do not yield "objective", experimental data. There are model and numerical approximations involved in all molecular calculations and if the results of the calculations are to be used wisely the limitations of the calculations must be borne in mind as much as the final results.

It is the purpose of this book to set out, in as direct a manner as possible, the theory and practice of orbital basis calculations so that the reader can evaluate the results of such calculations or, better, can program his own calculations in full knowledge of the approximations used.

It is assumed that the reader has some familiarity with the qualitative ideas of quantum chemistry at undergraduate level. This familiarity is assumed not for specific concepts or mathematical formulae but for general background and motivation. In the strict formal sense, very little previous knowledge is assumed (except in the short section on symmetry orbitals) and most of the necessary theory is developed as it is needed but a reader with no previous exposure to quantum chemistry would find the going a little thin.

The approach throughout is to get to the centre of the practical problems as quickly as possible - this means that many ideas are not developed in full generality. A list of recommended reading is given at appropriate points to guide the reader into related interesting areas. The references given in these reading lists are chosen on personal grounds rather than on historical importance or precedent. In keeping with the general aims of the work the references contain the material in a computationally convenient form or are in line with the notation used here.

Where it is relevant, program fragments are given which illustrate the implementation of the ideas discussed in the text. These fragments are coded in FORTRAN since this is still the most popular scientific high-level language. In every case an English equivalent is given in the text and following the two presentations will help the reader who is inexperienced in programming.

Throughout all stages of the development of the theory of orbital basis calculations the ideas are applied to a specific example - linear BeH_2. This system was chosen because it illustrates *all* the ideas of the general molecular calculation and yet is sufficiently small to enable all the data and results to be conveniently listed. A treatment of a large molecule such as ferrocene would show no features different *in principle* from the BeH_2 example.

Readers familiar with the notation of quantum chemistry may find it surprising that no use of the Dirac "bra-ket" notation is made. There are several reasons for this decision: in practical applications it is useful to display rather than disguise the explicit dependence of the integrand on the integrated variables; there is a very real danger of confusion between the "Dirac" notation for electron repulsion integrals and the more usual "charge-cloud" notation; and finally the use of arbitrary *functions* (not eigenvalues) inside the bra-ket symbol is not Dirac notation.

It is an axiom of computer programming that "there is no such thing as a completely de-bugged program" and, while every effort has been made to ensure that the program fragments used in the text are correct, errors may still remain if only because some of the fragments cannot be tested as they stand. Whenever the fragment could be tested it has been and, even though one man's test data is another man's trivial example, the BeH_2 data was used. The sample programs listed in Appendix C are in current use and can, therefore, be assumed to be correct.

Sheffield 1973 D.B.C.

ACKNOWLEDGEMENTS

It is a pleasure to thank my colleagues Mike Elder and Peter Dacre for useful discussions throughout the preparation of the manuscript; in particular Mike Elder heroically read much of the material in draft form - in constant danger of apoplexy brought on by my spelling. The reader familiar with quantum chemistry will recognise my debt to the methods and notation of Roy McWeeny. Finally, the photolitho reproduction process enables a full appreciation of Sue Rogers' typing to be made.

CONTENTS

1 INTRODUCTION

It is a striking paradox that quantum chemists, who in many universities and research institutes are the major users of computing facilities, are seeking essentially *qualitative* results. Their gigantic computer investigations into molecular electronic structure are largely the use of quantitative methods to obtain insight into the qualitative concepts used by experimental chemists. Of course, any investigation of qualitative ideas of "bonding", "valence" etc. is made much more precise and convincing by the successful prediction of molecular properties or the computation of electron densities which agree with diffraction data. However, the quantum chemist's plan is "to find out what electrons are doing in molecules" and the computation of molecular parameters is, in the main, a *source of evidence* rather than an end in itself.

In the strict terminology of quantum theory many of the chemists concepts are not "observables"; there is no hope of isolating a double bond, a localised molecular orbital, a bond polarity or even a bond energy. Nevertheless, much of chemistry and biochemistry is expressed in terms of these concepts and they have their own historical and methodological justification *in practice* - in any case they ante-date quantum theory and can be used autonomously. In order to maintain contact with and to influence the main stream of chemical thinking quantum chemistry must apply itself to the analysis, criticism and quantitative investigation of these concepts within the rigorous discipline of quantum theory. The construction of well-defined theoretical models which embody the ideas and experience of chemists enables a thoroughgoing examination of the strengths and weaknesses of the underlying assumptions to be made.

In addition to this "interpretative" function quantitative valence calculations have an important and decisive role to play when it is beyond the wit or technique of experimental workers to obtain certain critical data. The theories of chemistry often interpret observable phenomena in terms of the interaction of unobservable "effects" and frequently a phenomenon is a balance of two or more of these effects acting in opposition. In such a case almost any observation can be explained in terms of the relative strengths of the opposing effects and the whole explanation sounds rather hollow. It is in precisely these situations where the quantitative nature of computational methods can make a decisive contribution. A computation of the magnitudes of the various factors involved is much more convincing than the use of indirectly inferred quantities. Investigations of this kind enable the concepts and effects with a valid theoretical and physical background to be sorted from effects which are simply names.

In practice the efforts of the experimentalist and the computational worker are complementary. It is impossible for the experimental chemist to find out what forces are involved when (say) a protein molecule has its conformation forcibly changed at a particular site in one amino acid constituent. Contrariwise, the computational worker would quickly be swamped by numerical problems in any serious attempt to investigate (say) the effect of variations in ionic strength on the configuration of a protein. The experimental worker is, for the theoretician, a constant source of challenging problems and the quantum chemist can suggest fruitful lines of research to the experimentalist.

The various branches of spectroscopy present problems of a quite different character; here it is necessary to have a detailed knowledge of the energy-level structure of molecules, radicals and ions. The use of quantum mechanical calculations to interpret the interaction of radiation with matter over a large range of energy presents computational quantum chemistry with its severest quantitative test. There is a wealth of detailed experimental knowledge about the excited states of molecular systems ranging from X-ray transitions to NMR nuclear couplings. Attempts have been made to use molecular wave-functions to compute the transition energies of all of these spectra with differing degrees of success. The main result is a sound qualitative understanding of the dependence of the transition energies on molecular parameters.

Finally the most ambitious, and to chemists the most valuable, project: the computation of reaction paths. The difficulties here are ones of scale; the computation of the total energy of a composite system of two or more fragments in a large number of configurations. The only feasible study which can be made by the methods outlined in the following chapters is that of gas phase reactions between simple molecules which constitutes a very small part of chemistry. Solvents and catalysts play such an important role in the vast bulk of chemistry and biochemistry that, at the present time, the only contribution to this field we can sensibly make is to wish experimental workers luck in developing empirical theories.

No attempt is made in the rest of this work to pursue particular applications: the *methods* of quantum chemical calculations are developed.

SUGGESTIONS FOR FURTHER READING

"Nature of Quantum Chemistry" P.-O. Löwdin in Int.J.Quant.Chem., 1, 7 (1967)

2 THE SCHRÖDINGER EQUATION

2.1 NOTATION AND DEFINITIONS

The relation between the requirements of quantum theory and the concepts and reasoning of chemistry are best illustrated by consideration of what is involved in the formal solution of the known equations for the electronic structure of molecular systems. For a system of n electrons we assume that a typical electron i can be described by its position in space - (x_i, y_i, z_i) in some co-ordinate frame - and the value of its internal degree of freedom s_i ("spin" - s_i may be $\frac{1}{2}$ or $-\frac{1}{2}$). We write these four "co-ordinates" as $x_i = (x_i, y_i, z_i, s_i)$ and, when necessary, the three spatial co-ordinates (x_i, y_i, z_i) as r_i. Quantum theory ensures that all knowledge of the properties of a system of n electrons having these co-ordinates is contained in a function $\Psi(x_1,x_2,...,x_n)$ in the sense that

$$|\Psi(x_1,x_2,...,x_n)|^2 dx_1 dx_2 ... dx_n$$

is the probability that electron 1 has co-ordinates x_1, electron 2 simultaneously has co-ordinates x_2 etc. The function $|\Psi(x_1,x_2,...,x_n)|^2$ is thus the *probability density* for the electron distribution in a 4n dimensional "space".

The function $\Psi(x_1,x_2,...,x_n)$ is called the *molecular electronic wave function* and, for a particular state of the electronic system, is determined by the simultaneous solution of at least the following two equations:

$$\hat{H}\Psi(x_1,x_2,\ldots,x_n) = E\Psi(x_1,x_2,\ldots,x_n) \qquad (2.1.1)$$

$$\hat{P}\Psi(x_1,x_2,\ldots,x_n) = (-1)^P\,\Psi(x_1,x_2,\ldots,x_n) \qquad (2.1.2)$$

The first of these equations is the time-independent Schrödinger equation (or simply the Schrödinger equation) E is the total energy of the electronic system - the so-called eigenvalue of equation (2.1.1) - and Ψ is the required function. The operator $\hat{H}$ is discussed in detail in the following section. The second equation is a mathematical formulation of the Pauli principle for electrons. Both (2.1.1) and (2.1.2) are *eigenvalue* equations, the *operators* $\hat{H}$ and $\hat{P}$, when applied to the function $\Psi(x_1,x_2,\ldots,x_n)$, give a numerical multiple of the function. Standard nomenclature for these equations is to describe $\Psi(x_1,x_2,\ldots,x_n)$ and E as the eigenfunction and eigenvalue respectively of the operator $\hat{H}$. The "hat" over $\hat{H}$ and $\hat{P}$ emphasises their operator nature. Throughout this work we shall be concerned with the determination of an approximate $\Psi(x_1,x_2,\ldots,x_n)$ for the lowest state of the system - the one with lowest E.

2.2 THE SCHRÖDINGER EQUATION

Specification of the operator $\hat{H}$ in (2.1.1) defines the Schrödinger equation for the system under consideration. The operator $\hat{H}$ - the Hamiltonian operator or simply the "Hamiltonian" - has the following form for n electrons moving in the electrostatic field of N nuclei:

$$\hat{H} = \sum_{i=1}^{n} \hat{h}(i) + \sum_{i>j=1}^{n} \hat{g}(i,j) \qquad (2.2.1)$$

where

$$\hat{h}(i) = -\frac{\hbar^2}{2m}\nabla^2(i) + \sum_{\alpha=1}^{N} -\frac{Z_\alpha e^2}{\kappa_0 r_{\alpha i}} \qquad (2.2.2)$$

and

$$\hat{g}(i,j) = \frac{e^2}{\kappa_0 r_{ij}} \qquad (2.2.3)$$

($Z_\alpha e$ is the charge on a typical nucleus).

This definition contains some extreme contractions of notation and some physical approximations, all of which must be made clear in order for the mathematics and physics of the following sections to be comprehensible.

The quantities e, m and $\hbar$ are, respectively, the charge on the proton, the mass of the electron and Planck's constant divided by 2π. It is the rule in quantum chemistry to take these quantities as *units* of charge, mass and angular momentum. With this choice a coherent system of units - *atomic units* - can be developed, which facilitates the "book-keeping of constants" in computational work. The theoretical work is carried out entirely in atomic units and comparison with experiment is achieved by a single conversion at the end of the calculations. Definitions and values of atomic units and relevant conversion factors are given at the end of this chapter. Using κ_0 as a unit of permittivity further simplifies the equations.

In atomic units then

$$\hat{h}(i) = -\tfrac{1}{2}\nabla^2(i) + \sum_{\alpha=1}^{N} -\frac{Z_\alpha}{r_{\alpha i}} \qquad (2.2.2)$$

$$\hat{g}(i,j) = \frac{1}{r_{ij}} \qquad (2.2.3)$$

The notation $\hat{h}(i)$ is used to mean that the operator $\hat{h}$ is an operator *on the co-ordinates of electron i*, i.e. on only some or all of (x_i, y_i, z_i, s_i). Similarly $\hat{g}(i,j)$ operates only *on the co-ordinates of electrons i and j*. The explicit forms of the operators determine the precise nature of the operations of course.

The first term in (2.2.2) - the quantum mechanical kinetic energy operator - is essentially the Laplacian operator ∇^2. Again, the notation $\nabla^2(i)$ means that the Laplacian is applied to the co-ordinates of electron i, so that

$$-\tfrac{1}{2}\nabla^2(i) = -\tfrac{1}{2}\left(\frac{\partial^2}{\partial x_i^{\,2}} + \frac{\partial^2}{\partial y_i^{\,2}} + \frac{\partial^2}{\partial z_i^{\,2}}\right)$$

The second term of (2.2.2) is simply the electrostatic attraction energy of a unit negative charge in the field of N positive charges (typical charge Z_α) - the attraction energy between an electron and the nuclei. The distance from the electron to a typical nucleus is:

$$\begin{aligned} r_{\alpha i} &= |R_\alpha - r_i| \\ &= [(X_\alpha - x_i)^2 + (Y_\alpha - y_i)^2 + (Z_\alpha - z_i)^2]^{\frac{1}{2}} \end{aligned}$$

The operator $\hat{g}(i,j)$ defined by (2.2.3) is the Coulomb repulsion energy between two unit charges - the electron repulsion energy in atomic units. Again,

$$\begin{aligned} r_{ij} &= |r_i - r_j| \\ &= [(x_i - x_j)^2 + (y_i - y_j)^2 + (z_i - z_j)^2]^{\frac{1}{2}} \end{aligned}$$

is the separation between the charges. The operators $\hat{h}(i)$ and $\hat{g}(i,j)$ occur throughout quantum chemistry and are often called the "one-electron Hamiltonian" and "the electron repulsion operator" respectively. It is the differential form of $\nabla^2(i)$ which gives $\hat{H}$ its operator character; the electrostatic terms are simply multiplicative. The second summation in (2.2.1) denotes summation over all the distinct pairs (i,j) excluding the terms with i=j, that is, over all i and j provided i>j.

In writing the Schrödinger equation in the form (2.1.1) we have made the implicit assumption that the co-ordinates of the nuclei, although occurring in the specification of $\hat{H}$, *do not occur in the solutions* $\Psi(x_1,x_2,\ldots,x_n)$. Clearly this can not be the case, since changing the values of the R_α must affect $\hat{H}$ and therefore $\Psi(x_1,x_2,\ldots,x_n)$. What we *do* assume is that the dependence of Ψ on the nuclear co-ordinates is of a parametric type: we assume a fixed nuclear geometry and calculate Ψ for this fixed $\hat{H}$. With this in mind we have suppressed the appearance of the variables R_α in Ψ. Our assumption is equivalent to the physical idea that the electrons can adjust their motions instantaneously to any nuclear motion and is the basis of the Born-Oppenheimer approximation. *We shall always work within this fixed-nucleus approximation.*

An important objection to the form (2.2.1) of the electronic Hamiltonian operator is that it is manifestly not complete - there are no terms representing electron spin-orbit interactions or electron-nuclear spin coupling etc. The omission of *magnetic* terms from $\hat{H}$ is deliberate since magnetic effects are on a very much smaller energy scale than electrostatic effects. The inclusion of spin terms in the molecular Hamiltonian would involve a prohibitive increase in the computational complexity of the Schrödinger equation and any magnetic effects would certainly be swamped by the model approximations which we shall make in Chapter 3. We assume that the electronic distribution is determined by the forces described by the operator $\hat{H}$ in (2.2.1) - the so-called non-relativistic Hamiltonian. This latter approximation would, at first sight, suggest that we can omit all considerations of electron spin from the problem and concentrate on the co-ordinates r_i rather than x_i. This is true from the point of view of the Schrödinger equation (2.1.1) but the fact that equations (2.1.1) and (2.1.2) must have *simultaneous solutions* involves electron spin in an essential way in the molecular wave function.

2.3 THE PAULI PRINCIPLE

The Pauli principle, in its most general form for electrons, states:

> The wave function for a many electron system must be anti-symmetric with respect to exchange of the co-ordinates x_i (space and spin) of any two electrons.

Equation (2.1.2) is an operator form of this principle

$$\hat{P}\Psi(x_1,x_2,\ldots,x_n) = (-1)^p \Psi(x_1,x_2,\ldots,x_n)$$

$\hat{P}$ is an operator which produces an arbitrary permutation among the $x_1,x_2,\ldots,x_n$ (the space and spin variables of the electrons) and p is the *parity* of the permutation - the number of pair transpositions to which the permutation is equivalent. Then $(-1)^p$ is +1 or -1 corresponding to even or odd parity. This operator form is equivalent to the statement of the Pauli principle given above. Since the Pauli principle is a statement about permutations of the x_i, i.e. involves *all four* of the electron's co-ordinates, we must retain the spin variables in the function Ψ. In practice we shall be looking for *approximate* solutions of (2.1.1) which satisfy (2.1.2). It is very easy to find solutions to (2.1.2) - they can be simply written down - so we shall regard equation (2.1.2) as a *constraint* which any trial function must satisfy before being submitted to Schrödinger's equation. This constraint can be satisfied quite easily by choosing a *specific form* for any function with which we hope to make an approximation to the true solution of the Schrödinger equation.

2.4 CONSTRAINTS ON THE SCHRÖDINGER EQUATION

Equations (2.1.1) and (2.1.2) are the only two equations which every molecular wave function must satisfy; but there are other equations, which can be formulated as constraints on (2.1.1), which apply in particular cases. All these further constraints are applied to ensure that a molecular wave function or, more importantly an *approximate* molecular wave function, shall exhibit the correct symmetry properties of the molecular system which it represents.

It is usual to insist that the function Ψ represent a definite state of electron spin angular momentum:

$$\hat{S}^2 \Psi = S(S+1) \Psi$$

$$\hat{S}_z \Psi = M_S \Psi$$

where

$$\hat{S} = \sum_{i=1}^{n} \hat{s}(i)$$

and S, M_S are the total angular momentum and the z-component of angular momentum quantum numbers respectively. The operators $\hat{s}(i)$ are the spin angular momentum operators for electrons, which we shall discuss in connection with the definition of spin-orbitals in Chapter 3.

Many molecules have planes or axes of spatial symmetry. These symmetry elements mean that the electron distribution must be the same at equivalent points in the molecule, and this fact imposes constraints on the molecular wave functions. The consequences of these symmetry properties are discussed in Chapter 11. It is worth remarking at this point that the exact solution of (2.1.1)/(2.1.2) would automatically have the correct symmetry properties. These symmetry properties only appear as constraints on an *approximate* wave function and partially reflect the inadequacy of the form of the approximating function.

2.5 PROPERTIES OF THE MOLECULAR WAVE FUNCTION

The interpretation of $|\Psi|^2$ as a probability density means that all molecular wave functions must satisfy certain conditions in order that $|\Psi|^2$ have a physically reasonable form.

(i) The n electrons whose distribution is given by $|\Psi|^2$ must be wholly contained in the 4n-dimensional space so that the sum of all possible elements

$|\Psi(x_1,x_2,\ldots,x_n)|^2 dx_1 dx_2 \ldots dx_n$ must be finite. That is:

$$\int dx_1 \int dx_2 \ldots \int dx_n |\Psi(x_1,x_2,\ldots,x_n)|^2 \qquad (2.5.1)$$

must be finite. In fact it is conventional to scale Ψ by a numerical factor to *normalise* expression (2.5.1) to unity. When this is done the function Ψ is said to be normalised.

(ii) The probability density must clearly only have one value for one point $(x_1,x_2,\ldots,x_n)$ in the space. Thus $\Psi(x_1,x_2,\ldots,x_n)$ must be *single-valued* for all its variables.

(iii) The probability density must vary smoothly. The function Ψ must be a continuous function of the spatial co-ordinates $r_1,r_2,\ldots,r_n$. As we shall see, the dependence of Ψ on the spin "co-ordinates" is via a discontinuous label not a true function.

Any approximations to the molecular electronic wave function will obviously be the better for satisfying these three conditions.

2.6 DENSITY FUNCTIONS

From a chemical point of view much of the information contained in the function $|\Psi|^2$ is redundant; chemists rarely ask about the detailed, correlated, motions of the electrons in a molecule. Most chemical questions can be answered by a knowledge of the distribution of electrons in ordinary space and the *relative* distributions of the electrons in pairs, threes , etc. is a less important consideration.

The electron density - $\rho_1(x)$ say - is easily defined as a function of the four co-ordinates of a single electron by integration over the possible positions of the other electrons:

$$\rho_1(x) = n \int dx_2 \int dx_3 \ldots \int dx_n |\Psi(x, x_2,\ldots,x_n)|^2 \qquad (2.6.1)$$

where we have dropped the suffix from the remaining variable. The factor n in (2.6.1) arises because there are n identical contributions to $\rho_1(x)$ coming from the n possible choices of non-integrated co-ordinate. This function can be interpreted as the probability density of finding *any* electron at r with spin s. Using a continuous model for the electron probability density it is the total charge/spin density.

Another density function - the pair distribution function - is useful in the interpretation of electronic properties which depend on the correlated motions of electron pairs. It is defined in a similar way to $\rho_1(x)$, integrating the total distribution over all but two sets of co-ordinates and summing the n(n-1) identical contributions.

$$\rho_2(x_1,x_2) = n(n-1) \int dx_3 \int dx_4 \dots \int dx_n |\Psi(x_1,x_2,\dots,x_n)|^2 \quad (2.6.2)$$

This density function is a measure of the probability of finding an electron at x_1 and another electron simultaneously at x_2 whatever the positions of the remaining n-2 electrons. From the definitions (2.6.1) and (2.6.2) of ρ_1 and ρ_2, or from the physical interpretation, we must have

$$\int dx_2 \rho_2(x,x_2) = (n-1)\rho_1(x)$$

and (2.6.3)

$$\int dx \rho_1(x) = n$$

These density functions depend on the *four* co-ordinates (x,y,z,s) of the electrons and so the densities they describe are not quite the electron density of diffraction experiments or the idea which springs to mind when electron density is discussed in a valence theory context. It is therefore useful to define a function of ordinary three-dimensional space which has a direct interpretation as the charge density in a molecule. Integrating over the spin co-ordinate of $\rho_1(x)$ gives the required function:

$$P(r) = \int ds \rho_1(x) \quad (2.6.4)$$

P(r) contains much of the qualitative and quantitative chemical information of $\Psi(x_1,x_2,\ldots,x_n)$ in a more readily digestible form. Combining (2.6.1) and (2.6.4) we have the full definition:

$$P(r) = n\int ds\int dx_2\ldots\int dx_n|\Psi(x,x_2,\ldots,x_n)|^2 \qquad (2.6.5)$$

Much of quantum chemistry consists of an analysis and interpretation of changes in electron density on bonding. We shall therefore be submitting P(r) to rather close scrutiny in later chapters.

2.7 SOLUTIONS OF THE SCHRÖDINGER EQUATION

Inspection of the form of the molecular Hamiltonian (2.2.1) shows that the actual analytical problem presented by the solution of the Schrödinger equation is a partial differential equation in 3n dimensions. The existence of the operators $\hat{g}(i,j)$ means that the problem is *essentially* one of 3n dimensions: in general no further reduction into equations of smaller dimension is possible. The Schrödinger equation for benzene is thus a partial differential equation in 126 dimensions! For one-electron systems (n=1) there are, of course, no electron repulsion terms $\hat{g}(i,j)$ and the resulting three-dimensional equation can be solved. This can be done by classical analytical methods for N=1,2 (one-electron atoms and diatomics) and numerically for N>2 (one-electron polyatomic molecules). The numerical solution of the Schrödinger equation for n>2 and N>1 presents extreme computational difficulties. In fact, it has been clear for some years now that there is no hope of obtaining "exact" solutions of the Schrödinger equation for molecules.

From the chemist's point of view the insolubility of the Schrödinger equation is rather irrelevant since, even if the numerical solutions were available, they would be huge lists of numbers - values of $\Psi(x_1,x_2,\ldots,x_n)$ at chosen values of $(x_1,x_2,\ldots,x_n)$ perhaps stored on a magnetic tape or disk. The chemical information would be present in Ψ but in an extremely inconvenient, redundant and inaccessible form. The Schrödinger equation is simply an equation for the motion of n electrons in the potential field of each other and N nuclei; it does not "know" that this system is a molecule or that molecules are considered by chemists to be made of atoms. In short, the concepts and ideas of chemistry are not contained in the Schrödinger equation. Thus, if we rely on the purely mathematical approach, every molecule would have to be treated as a new problem; methane as a ten-electron tetrahedral system and ethane as an eighteen-electron system. The idea of an approximately constant C-H fragment would only be revealed after considerable analysis of the electron densities given by the two molecular wave functions.

More importantly, the most fundamental concept of chemistry - the idea of *valence* - cannot be obtained from a molecular wave function since valence involves a *comparison* of two electron densities. The electron density *change* on going from the seperate atoms to the molecule is basic to any valence interpretation of molecular structure. Thus, for methane, we would have to solve the Schrödinger equation for methane, carbon and hydrogen; compute the separate electron densities from each wave function and examine the change in electron density:

$$P_{CH_4}(r) - P_C(r) - P_{H_1}(r) - P_{H_2}(r) - P_{H_3}(r) - P_{H_4}(r).$$

This change has to be compared with the results of a similar calculation on carbon, hydrogen and ethane:

$$P_{C_2H_6}(r) - P_{C_1}(r) - P_{C_2}(r) - \sum_{i=1}^{6} P_{H_i}(r)$$

All this to obtain a quantitative test of the invariance of a C-H bond: a single chemical concept from a vast computation. Thus from a quantum chemist's point of view the route to chemical information would be too complex even if the solutions of the Schrödinger equation could be computed and stored. Almost all the questions of chemistry can be answered with only a fraction of the information contained in Ψ, the molecular wave function. This fact shows up even at the most elementary qualitative level. Most introductory courses on valence theory begin by considering the bonding in H_2^+ (or H_2) and yet make no use of the known solutions of the Schrödinger equation for H_2^+. The bonding is invariably discussed in terms of the wave functions of the separate atoms - introducing ideas of valence from the very start of the theory. As chemists we must *start* from the idea of valence - the atomic structure of molecules - rather than starting from a partial differential equation.

We can perhaps risk a gloss on the famous quote of P.A.M. Dirac:

> "The underlying physical laws necessary for the mathematical theory of a large part of Physics and the whole of Chemistry are thus completely known, and the difficulty is only that the exact application of these laws leads to equations much too complicated to be soluble."

Even if the equations were soluble we would be faced with the task of extracting the information. Where would we find aromaticity in a function of 126 variables for benzene? Chemistry has developed its own concepts and some of these concepts are of proven value and cannot be discarded because of scientific discoveries at a "lower level" any more than biological concepts are undermined by chemical knowledge. *As far as possible* chemical problems should have chemical answers. To a chemist, the "explanation" of phenomena offered by the solution of the Schrödinger equation would seem to be "a restatement of the problem in more technical language". To say to a chemist that the "reason" why conformation A of a molecule is more stable than conformation B is that A has a lower energy than B is not just tautologous - it's irritating.

The arguments presented in this section have been aimed at showing that quantum chemistry and quantum chemical calculations can claim a certain *autonomy* if they attempt to pose chemical questions to the Schrödinger equation in a way which yields chemically useful information. The methods developed in the following chapters aim to do just this, and would not therefore be made obsolete by the discovery of methods of solving the Schrödinger equation exactly for molecules.

2.8 ATOMIC UNITS

The conventions we adopted in 2.2 define the natural units for molecular computational work. The atomic units of mass, charge, angular momentum (or "action", dimensions MLT^{-1}) and permittivity enable all other quantities to be expressed in a coherent system of atomic units (a.u.). The derived units relevant to this work are those of length and energy. The Table below gives names and numerical values (in SI units) of the most common atomic units met in molecular computational work.

Physical Quantity	Atomic Unit	Value in SI
mass	electron mass	9.1091×10^{-31} kg
charge	proton charge	1.6021×10^{-19} C
action	Planck's Constant/2π	1.0545×10^{-34} Js
permittivity	κ_0 (= $4\pi\varepsilon_0$)	1.1127×10^{-10} Fm^{-1}
length	radius of 1st Bohr orbit	0.52917×10^{-11} m
energy	twice the ground state energy of hydrogen atom	4.3594×10^{-18} J

The units of length and energy are commonly called the Bohr and the Hartree respectively. The unit of energy is particularly important in comparisons with experiment and the value of one a.u. of energy in various "hybrid" energy units is given below.

Units	eV	kJ mol^{-1}	kcal mol^{-1}	cm^{-1}
Value of 1 a.u.	27.211	2625.51	627.51	219475*

* using the H atom Rydberg constant

Throughout the following chapters the symbols appearing in the equations are understood to be the *numerical values of the quantities* in atomic units. That is, we specifically avoid the use of the "symbol = quantity" notation in which all physical quantities are divided by the units. Our equations are complex enough without the use of additional cumbersome nomenclature: specifically we will write "r" not "r/(a.u.)" or "r/a_0" (a_0 = 1 Bohr).

SUGGESTIONS FOR FURTHER READING

"The Fundamental Principles of Quantum Mechanics" by E.C. Kemble (Dover 1958) contains a rather daunting discussion of the properties of the solutions of the many-particle Schrödinger Equation in Chapter 6.

"Some Recent Advances in Density Matrix Theory" by R. McWeeny in Rev.Mod.Phys., 32 235 (1960) discusses the density function and its use in Quantum Chemistry. A general discussion of Atomic Units is given in Appendix 1 of "Quantum Mechanics: Methods and Basic Applications" by R. McWeeny (Pergamon 1973).

3 THE ORBITAL APPROXIMATION

3.1 THE ORBITAL MODEL

In Section 2.7 we gave reasons why we do not wish to approach the solution of the Schrödinger equation as an exercise in partial differential equations. Rather, we will make use of chemical and physical knowledge in formulating a method of solution. Our approach is, basically, to use larger units than the differential elements of a 3n-dimensional space to build up approximate solutions. We can therefore expect that, just as an architect who designs in pre-fabricated units has less flexibility than one who uses bricks, perhaps some of the finer details of the molecular electronic structure will evade our analysis. In these cases we must be prepared to extend or amend our model to try to give a satisfactory description.

Since the development of the first qualitative ideas about the structure of atoms and molecules - in terms of the wave functions of the hydrogen atom - the language of chemistry has been overwhelmingly an "orbital" language. The basic units of valence theory are the hydrogen atom wave functions - atomic orbitals. The combination of atomic wave functions to form molecular wave functions is the backbone of interpretative quantum chemistry.

3.2 THE SCHRÖDINGER EQUATION IN AN ORBITAL BASIS

The term *orbital* has a rather imprecise meaning in quantum chemistry, particularly in computational usage; originally it meant

> The solution of any real or model single-electron Schrödinger Equation

but current usage is more vague and can best be summarised as

> Any continuous function of three dimensions whose square is integrable.

The latter definition includes the former, of course, and the formal properties of the second definition are all that are required for the mathematical methods given later. However, as we shall see, the general physical interpretation of molecular wave functions is biased toward the former definition, particularly when atomic orbitals are used.

There are two general ways of using chemical and physical information in looking for approximate solutions of the Schrödinger equation: model approximations and numerical approximations. As the name suggests, "model" approximations consist of making simplifying assumptions about the nature of the physical system under investigation. Mathematically, this means restricting the *form* of any function with which we hope to approximate the molecular wave function. We must seek a mathematical form which embodies a model based on our chemical intuitions about molecular electronic structure. Numerical approximations, on the other hand, are made at a lower level - they pre-suppose that a model approximation has been made. They involve the use of experimentally observed (or inferred) magnitudes for the various energy integrals in place of the theoretical expressions for these quantities. Roughly speaking, model approximations employ chemical and physical knowledge and numerical approximations circumvent computational difficulties. We shall be concerned entirely with model approximations.

The way the molecular Hamiltonian is written in (2.2.1)

suggests a general model approach to the approximate solution of the associated Schrödinger equation

$$\hat{H} = \sum_{i=1}^{n} \hat{h}(i) + \sum_{i>j=1}^{n} \hat{g}(i,j)$$

If we neglect the second set of terms, we are left with the simpler Schrödinger equation:

$$H'\Psi' = (\sum_{i=1}^{n} \hat{h}(i))\Psi' = E\,\Psi' \qquad (3.2.1)$$

where the prime has been added to Ψ and $\hat{H}$ to denote the simplification. This equation can be broken down into n separate equations

$$\hat{h}(i)\mu_j(r_i) = \varepsilon_j\mu_j(r_i) \qquad (3.2.2)$$

$$i = 1,2,\ldots n$$
$$j = 1,2,\ldots\infty$$

Here, the $\mu_j(r_i)$ and ε_j are the eigenfunctions and eigenvalues of $\hat{h}(i)$. Since the $\hat{h}(i)$ only differ by the numbering of the variables all n equations (3.2.2) are identical. It is easy to verify by direct substitution into (3.2.1) that the required eigenfunctions and eigenvalues of the n-electron problem are

$$\Psi = \mu_{j_1}(r_1)\mu_{j_2}(r_2)\mu_{j_3}(r_3)\ \ldots\ \mu_{j_n}(r_n) \qquad (3.2.3)$$

and

$$E = (\varepsilon_{j_1} + \varepsilon_{j_2} + \varepsilon_{j_3}\ \ldots + \varepsilon_{j_n}) \qquad (3.2.4)$$

(j_1, j_2, etc. represent a selection of the possible solutions of (3.2.2)).

We simply have to remember that $\hat{h}(i)$ only operates on functions of r_i:

$$\hat{h}(i)\mu_{j_1}(r_1) \ldots \mu_{j_i}(r_i) \ldots \mu_{j_n}(r_n)$$

$$= \mu_{j_1}(r_1) \ldots \varepsilon_{j_i}\mu_{j_i}(r_i) \ldots \mu_{j_n}(r_n).$$

In these expressions the functions $\mu_j(r)$ are orbitals in the sense of both definitions given at the beginning of this section. The fact that the true molecular Hamiltonian does contain electron repulsion terms $\hat{g}(i,j)$ means that the function (3.2.3) is not the true molecular wave function. Examination of equation (3.2.4) shows that it is only the *sum* of the ε_j which is important in defining E; therefore the product of *any* n orbitals $\mu_j(r)$ is a satisfactory solution of (3.2.1) provided that the associated ε_j add up to E.

If Ψ_1' and Ψ_2' are two such constructions and D_1 and D_2 are numerical coefficients, then

$$\hat{H}'(D_1\Psi_1' + D_2\Psi_2') = E(D_1\Psi_1' + D_2\Psi_2') \qquad (3.2.5)$$

Any linear combination of such functions is also an eigenfunction of (3.2.1).

When the full Hamiltonian - including $\hat{g}(i,j)$ - is considered, these functions are not eigenfunctions. However, if we can choose the orbitals $\mu_j(r)$ and/or the coefficients D_i carefully enough to compensate for the effect of the $\hat{g}(i,j)$, we can hope that this orbital model will present a systematic approach to quantum chemistry:

> *If it is possible to partition, in some approximate way, the n-electron Hamiltonian into n separate one-electron Hamiltonians, then an approximate solution of the full Schrödinger equation is given by a linear combination of products of the orbitals defined by the one-electron Schrödinger equations.*

The choice of the particular method of partitioning the Hamiltonian defines the orbitals $\mu_j(r)$.

This approximation method of seeking solutions to the Schrödinger equation defines a whole class of *orbital approximations*. The choice of $\mu_j(r)$ through the method of partitioning $\hat{H}$ defines a physical model, and this point is taken up in detail for molecular systems in the following chapter. The best possible solution to the Schrödinger equation using this orbital model is the one for which the $\mu_j(r)$ are optimum and the linear coefficients D_i of (3.2.5) are chosen to approximate the true molecular wave function most closely. The choice of the orbitals $\mu_j(r)$ defines a set of *building blocks* for the molecular wave function, and the way in which the $\mu_j(r)$ appear in the optimum wave function gives an interpretation of Ψ in terms of the $\mu_j(r)$.

Physically, this means interpreting the distribution of n electrons in terms of the separate distributions of the model one-electron orbitals.

3.3 ELECTRON SPIN AND THE IMPOSITION OF THE PAULI PRINCIPLE

It is immediately evident from an examination of the product form of Ψ' in (3.2.3) that the approximation function does not satisfy the Pauli principle. The spin coordinate is not mentioned at all in the $\mu_j(r)$. The most direct way of introducing electron spin into our orbital theory is to introduce two spin "functions" which formally describe the two possible values of the electron's internal degree of freedom. The spin factor α is added to the definition of an orbital to mean that the electron with this spatial distribution has spin angular momentum $\frac{1}{2}$. The factor β means the electron has "opposite spin" - spin angular momentum $-\frac{1}{2}$. In order to comply with the formalism of angular momentum theory, these functions are regarded as eigenfunctions of the spin operators $\hat{s}^2$ and $\hat{s}_z$ which define the total spin angular momentum and the z-component of the spin angular momentum:

$$\hat{s}^2\alpha = \tfrac{1}{2}(\tfrac{1}{2}+1)\alpha$$

$$\hat{s}^2\beta = \tfrac{1}{2}(\tfrac{1}{2}+1)\beta$$

$$s_z\alpha = \tfrac{1}{2}\alpha \ ; \qquad s_z\beta = -\tfrac{1}{2}\beta$$

Thus, forming the product of a set of spatial orbitals $\{\mu_j\}$ with the two possible spin factors defines a set of so-called *spin orbitals* $\{\lambda_j\}$ (say) where each member of $\{\lambda_j\}$ is a member of $\{\mu_j\}$ multiplied by α or β. This procedure gives twice as many spin orbitals as spatial orbitals:

$$\lambda_k(x) = \lambda_k(r,s) = \mu_j(r)\alpha(s)$$

$$\lambda_{k+1}(x) = \lambda_{k+1}(r,s) = \mu_j(r)\beta(s)$$

For many formal purposes it is useful to be able to speak of "an electron occupying spin orbital λ_k". If $\lambda_k = \mu_j\alpha$, for example, then this statement means the electron has a spatial distribution given by $|\mu_j|^2$, and has spin angular momentum $\tfrac{1}{2}$. It is also common to speak of the two possible values of the spin coordinate as "spin-up" and "spin-down".

We can now re-write equation (3.2.3) substituting λ's for μ's and note that the result is still an eigenfunction of (3.2.1) as the spatial operators $\hat{h}(i)$ have no effect on the spin functions. The function

$$\prod_{i=1}^{n} \lambda_i(x_i) \qquad (3.3.1)$$

is a function of all four coordinates of each electron. We can impose the Pauli principle simply by forming all possible permutations of the x_i and adding the results with a factor of $(-1)^p$ for each permutation. This ensures that the Pauli principle is obeyed independently of the nature of the λ_i. The operation of forming a function which is anti-symmetric with respect to permutations of the x_i can be neatly summarised in terms of the permutation operators $\hat{P}$:

$$\hat{A} = \sum_P (-1)^p \hat{P} \qquad (3.3.2)$$

where A is the *anti-symmetrising operator* and the summation ranges over all possible permutations.

A function which embodies our orbital model and satisfies the Pauli principle is thus given by

$$\Phi = \hat{A} \prod_{i=1}^{n} \lambda_i(x_i) \qquad (3.3.3)$$

The function Φ is still a solution of (3.2.1) since all the n! permutations are solutions of (3.2.1) with the same energy, differing only in the names of the variables. We can make Φ a little less forbidding by noting that it is a *determinant* whose rows are defined by the λ_i and whose columns are enumerated by the x_i

$$\Phi(x_1, x_2 \ldots x_n) = \begin{vmatrix} \lambda_1(x_1)\lambda_1(x_2) \ldots \lambda_1(x_n) \\ \lambda_2(x_1)\lambda_2(x_2) \ldots \lambda_2(x_n) \\ \lambda_3(x_1)\lambda_3(x_2) \ldots \lambda_3(x_n) \\ \cdots\cdots\cdots\cdots \\ \cdots\cdots\cdots\cdots \\ \lambda_n(x_1) \quad \ldots\ldots \lambda_n(x_n) \end{vmatrix} \qquad (3.3.4)$$

Equation (3.3.4) is just the historically established notation for (3.3.3). Note that if any λ_i appears more than once the function Φ is zero. Within the orbital model then, the Pauli principle reduces to "spin-orbitals can be, at most, singly occupied" or "spatial orbitals can contain either one electron, or two electrons with opposite spin". The converse is also true; so that if we *always work with anti-symmetric determinants of spin-orbitals* we can be sure that our approximate wave function satisfies the Pauli principle and we can concentrate our attention on the solution of the Schrödinger equation.

It is worthwhile, before continuing, to check that our determinantal functions satisfy the requirements of Section 2.5. Obviously the continuity and single-valued nature of the approximate n-electron wave function are dependent on the continuity and behaviour of the component orbitals. Since the orbitals are interpretable as solutions of a Schrödinger equation in their own right these conditions will normally be met. It only remains to show that our determinantal function can be normalised. Multiplying (3.3.4) by N and insisting that

$$N^2 \int dx_1 \int dx_2 \ldots \int dx_n |\Phi(x_1,x_2,\ldots x_n)|^2 = 1$$

defines N in terms of integrals over the orbitals. It is an elementary (but tedious) exercise in the algebra of determinants to show that

$$N = \left\{ \begin{vmatrix} 1 & S_{12} & S_{13} & \cdots & S_{1n} \\ S_{21} & 1 & S_{23} & & S_{2n} \\ \cdots & \cdots & \cdots & \cdots & \cdots \\ \cdots & \cdots & \cdots & \cdots & \cdots \\ S_{n1} & & & & 1 \end{vmatrix} \right\}^{-\frac{1}{2}} (n!)^{-\frac{1}{2}} \qquad (3.3.5)$$

where

$$S_{ij} = \int dx \lambda_i(x) \lambda_j(x) \qquad (3.3.6)$$

It is convenient to have a notation for the *normalised* anti-symmetrised product in terms of the component λ_i; we shall use

$$\det\{\lambda_1(x_1)\lambda_2(x_2) \ldots \lambda_n(x_n)\} \qquad (3.3.7)$$

to mean the normalised determinant (3.3.4) and reserve Φ to denote determinantal functions.

3.4 FULL STATEMENT OF THE ORBITAL MODEL

We can now restate the main conclusion of 3.2 together with the additional constraint of the Pauli principle:

> If a choice of physically reasonable spatial orbitals (in the sense of 3.2) is made and a set of associated spin-orbitals formed, then an approximate solution of the molecular Schrödinger equation is given by a linear combination of determinants of these spin-orbitals.

In mathematical notation

$$\Psi = \sum_i D_i \Phi_i \tag{3.4.1}$$

In formulating the solution of the Schrödinger equation in this way we have made an important shift of emphasis. In place of the attempt to calculate a function of 4n-dimensional space we are now looking for a set of functions of ordinary three dimensional space - the $\mu_i(r)$ - and some numerical coefficients - the D_i. The mass of sheer numerical data contained in Ψ is made more digestible by being absorbed into the *functional forms* of the $\mu_i(r)$. Indeed, it is to these μ_i that we must look for re-usable elements of molecular electronic structure; functions reflecting the recurring, almost environment independent, regions of electron density. We should perhaps note in passing that the Schrödinger equation cannot be solved by sleight of hand - our transformations of the form of the equation have shifted the bulk of the numerical work to another area which is more amenable to systematic computation. The most important *practical* problem to be solved is the optimisation of the orbitals μ_i and/or the linear coefficients D_i. Fortunately there is an all-embracing theorem which will enable both types of optimisation to be performed.

3.5 THE VARIATION METHOD

The Schrödinger equation, as given in (2.1.1), is not in a form which lends itself to the systematic use of approximation methods. In this section we re-cast the equation into an equivalent form which is particularly suited to computational implementation. In order to effect this transformation we must go a little more deeply into the formal properties of the solutions of the Schrödinger equation. Readers who are willing to accept the last (italicised) paragraph of this section without proof may, without serious loss, omit the rest of the section.

The operator $\hat{H}$ defined in 2.2 has the important symmetry property that, for any functions satisfying the conditions of 2.5 - $\Theta_1(x_1,x_2,\ldots x_n)$ and $\Theta_2(x_1,x_2,\ldots x_n)$, say -

$$\int d\tau \Theta_1^*(x_1,\ldots x_n)\hat{H}\Theta_2(x_1,\ldots x_n) = \int d\tau \{\hat{H}\Theta_1(x_1,\ldots x_n)\}^*\Theta_2(x_1,\ldots x_n)$$

(where the shorthand notation $d\tau$ has been used for $dx_1 dx_2 \ldots dx_n$). In particular, if the functions Θ_1 and Θ_2 are *real*

$$\int d\tau \Theta_1 \hat{H}\Theta_2 = \int d\tau \Theta_2 \hat{H}\Theta_1 \tag{3.5.1}$$

This *Hermitian symmetry* property enables us to establish some of the properties of the solutions of the Schrödinger equation associated with $\hat{H}$. If we look at the real wave functions of two different states of the same system - Ψ_κ and Ψ_λ, say - then:

$$\hat{H}\Psi_\kappa = E_\kappa \Psi_\kappa \tag{3.5.2}$$

$$\hat{H}\Psi_\lambda = E_\lambda \Psi_\lambda \tag{3.5.3}$$

where E_κ and E_λ are the relevant electronic energies. Multiplying (3.5.2) by Ψ_λ and (3.5.3) by Ψ_κ and integrating we obtain:

$$\int d\tau \Psi_\lambda \hat{H}\Psi_\kappa = E_\kappa \int \Psi_\lambda \Psi_\kappa d\tau$$

and

$$\int d\tau \Psi_\kappa \hat{H} \Psi_\lambda \;=\; E_\lambda \int \Psi_\kappa \Psi_\lambda d\tau$$

since E_κ and E_λ are simply numbers.

Obviously

$$\int d\tau \Psi_\lambda \Psi_\kappa \;=\; \int d\tau \Psi_\kappa \Psi_\lambda$$

and (3.5.1) ensures the equality of the two left-hand side expressions. Subtracting the two equations and re-arranging the result we have:

$$(E_\kappa - E_\lambda)\int d\tau \Psi_\lambda \Psi_\kappa \;=\; 0 \qquad (3.5.4)$$

when $\kappa=\lambda$, $E_\kappa=E_\lambda$ and we know that the integral in (3.5.4) is the normalisation integral for Ψ_κ. When $\kappa\neq\lambda$, $E_\kappa \neq E_\lambda$ and therefore

$$\int d\tau \Psi_\kappa \Psi_\lambda \;=\; 0 \qquad (3.5.5)$$

Wave functions associated with different energies of the same system are orthogonal (the product integrates to zero).

We have seen that it may always be arranged that Ψ_κ is normalised, so the normalisation and orthogonality conditions may be summarised neatly by use of the Kronecker delta:

$$\int d\tau \Psi_\kappa \Psi_\lambda \;=\; \delta_{\kappa\lambda} \;=\; \begin{cases} 1 & \kappa=\lambda \\ 0 & \kappa\neq\lambda \end{cases} \qquad (3.5.6)$$

In the case of so-called degenerate electronic states of a molecule there are several independent wave functions ($\Psi_{\kappa 1}$, $\Psi_{\kappa 2}$,...) having energy E_κ and (3.5.5) does not follow from (3.5.4). It is still possible, however, to choose the functions $\Psi_{\kappa 1}$, $\Psi_{\kappa 2}$,... such that (3.5.6) holds.

The whole set of functions which are solutions of a particular Schrödinger equation form a *complete set* of anti-symmetric functions in 3n variables. Any anti-symmetric, continuous, normalisable function can be expanded as a linear combination of the solutions of a Schrödinger equation of the same dimension.

Returning now to the solution of the Schrödinger equation

$$\hat{H}\,\Psi = E\,\Psi$$

we can multiply the equation from the left by Ψ and integrate, giving

$$\int d\tau\Psi\hat{H}\Psi = E\int d\tau\Psi\Psi$$

or

$$E = \int d\tau\Psi\hat{H}\Psi \Big/ \int d\tau\Psi\Psi \tag{3.5.7}$$

It is a surprising fact that if Ψ is replaced by *any approximate* function $\tilde{\Psi}$ in (3.5.7) then the value of the computed approximate energy expression ($\tilde{E}$ say) is always higher than the lowest true solution of the Schrödinger equation:

$$\tilde{E} = \frac{\int d\tau\tilde{\Psi}\,\hat{}\,\tilde{\Psi}}{\int d\tau\tilde{\Psi}\tilde{\Psi}} > E \tag{3.5.8}$$

The proof of this inequality is quite simple and illustrates some of the formal methods of quantum mechanics in manipulating functions whose *forms are unknown* but some of whose properties are known. We assume that the trial function $\tilde{\Psi}$ can be written as a linear expansion in terms of the (unknown) solutions of the Schrödinger equation:

$$\tilde{\Psi} = \sum_{\kappa} f_{\kappa}\,\Psi_{\kappa} \tag{3.5.9}$$

where the expansion length may be infinite. Substituting (3.5.9) into (3.5.8) and using the fact that

$$\int d\tau\Psi_{\kappa}\hat{H}\Psi_{\lambda} = E_{\lambda}\int d\tau\Psi_{\kappa}\Psi_{\lambda} = 0$$

we have

$$\tilde{E} = \sum_{\kappa} f_{\kappa}^2 \int d\tau \Psi_{\kappa} \hat{H} \Psi_{\kappa}$$

i.e.

$$\tilde{E} = \sum_{\kappa} f_{\kappa}^2 \, E_{\kappa} \qquad (3.5.10)$$

For simplicity we have assumed that $\tilde{\Psi}$ is normalised:

$$\sum_{\kappa} f_{\kappa}^2 = 1 \qquad (3.5.11)$$

since this can always be arranged.

Now multiplying (3.5.11) by E - the true energy - and subtracting from (3.5.10)

$$(\tilde{E} - E) = \sum_{\kappa} f_{\kappa}^2 (E_{\kappa} - E)$$

But, if E is the *lowest* energy, $(E_{\kappa} - E)$ is always non-negative and, of course, all the f_{κ}^2 are positive; therefore:

$$(\tilde{E} - E) > 0$$

or

$$\tilde{E} > E,$$

proving (3.5.8).

This important and very general result suggests a practical procedure for the optimisation of any trial molecular wave function, and in particular, the orbital model wave function developed in 3.4. The approximate model wave function is substituted into the variational expression (3.5.8) and the values of any adjustable parameters contained in the wave function - the forms of the orbitals and the linear coefficients D_i - are varied until a minimum is found in the expression (3.5.8). The values of the adjustable parameters at the minimum define the best possible wave function *of that particular functional form:* the best possible description of the molecular electron distribution consistent with the limitations of the model.

The proof of the variation theorem (3.5.8) given above has a rather unsatisfactory "feel" about it when encountered for the first time. It is not clear to the beginner that a result obtained by formally expanding the function $\tilde{\Psi}$ in terms of an unavailable complete set has any bearing on what happens in practice when, for example, an orbital parameter is varied in (3.5.8). It usually takes a little time to appreciate the consequences of these formal expansion methods, but the enormous advantages of being able to use the whole arsenal of Linear Algebra in quantum theory makes the effort worthwhile.

In transforming the Schrödinger equation from the differential equation form to the apparently equivalent variational form (3.5.8) we have unwittingly lost some of the finer points of the solution. A variational solution of (3.5.8) is obtained by minimising the value of an *integrated expression;* the solution is the best possible solution of the model type *in the mean*. The differential Schrödinger equation has *point by point* solutions. We can therefore expect that any variationally-determined approximate solution of the Schrodinger equation will not, in principle, give a good description of molecular properties which depend on the value of $\tilde{\Psi}$ at *particular points* in space. Molecular properties depending on various integrations of $\tilde{\Psi}$ should, however, be well reproduced. The most important "point properties" are spin hyperfine coupling constants.

In summary:

The adjustable parameters contained in a model wave function $\tilde{\Psi}$ are optimum when the energy expectation value for this function

$$\tilde{E} = \frac{\int d\tau \tilde{\Psi}\hat{H}\tilde{\Psi}}{\int d\tau \tilde{\Psi}\tilde{\Psi}}$$

is a minimum with respect to variations in those parameters. More concisely, the optimum wave function of a given functional form is determined by

$$\delta \tilde{E} = 0$$

This variation principle applies strictly to only the lowest state of a given symmetry and is most widely used to determine ground state wave functions.

3.6 USE OF THE VARIATION PRINCIPLE

The conclusions of the previous section have given us a technique for the computation of the approximate, orbital model wave function:

$$\Psi = \sum_k D_k \Phi_k \tag{3.6.1}$$

where

$$\Phi_k = \det\{\lambda_1(x_1)\lambda_2(x_2) \ldots \lambda_n(x_n)\}$$

and

$$\lambda_i(x) = \mu_j(r) \times (\alpha \text{ or } \beta)$$

We have dropped the tilde from above the Ψ since, from now on, we shall always be dealing with approximate wave functions. The expression (3.6.1) is substituted into the variational expression (3.5.8) which is then minimised with respect to the coefficients D_k and any parameters contained in the definition of the spatial orbitals $\mu_j(r)$. In practice, this "full" optimisation process is far too complex and time consuming for many-electron systems of chemical interest, and some restricted form of optimisation has to be carried out. The unrestricted optimisation of the $\mu_j(r)$ and the D_k is a "coupled" problem since obviously the values of the D_k are dependent on the forms of the μ_j and vice versa. This full optimisation of expression (3.6.1) is called the Multi-Configuration Self-Consistent Field method and will only be mentioned briefly in the closing sections of this work.

The word *configuration* is used to mean a particular choice of n occupied spin-orbitals forming a single term in (3.6.1).

Thus a configuration is a physical description of the mathematical "determinant of spin-orbitals" idea*. The two main model approximations of quantum chemistry will be derived and described in Chapter 5: they correspond to the optimisation of *either* the μ_j *or* the D_k. The most widely used method is to retain only a *single term* in (3.6.1) ($D_1 = 1$; $D_i = 0$; $i>1$) and throw all the computational effort into choosing the best possible *orbitals* in this single configuration. This is the approach of the Molecular Orbital model and many of the actual implementations we shall be studying are of the computational details of this model.

The other method which uses fixed orbitals μ_j and optimises the coefficients D_i of an *essentially multi-configuration* wave function is the Valence-Bond method which is used very widely by chemists in a qualitative way, but has found little quantitative application.

3.7 THE LINEAR VARIATION METHOD

The computation of optimum linear expansion coefficients - the D_i of (3.6.1) - when the functions Φ are not optimised, is particularly easy to formulate in general: if

$$\Psi = \sum_k D_k \Phi_k$$

then substitution in (3.5.8) gives

$$\tilde{E} = \sum_{i,j} D_i D_j H_{ij} \Big/ \sum_{i,j} D_i D_j S_{ij} \qquad (3.7.1)$$

where

$$H_{ij} = \int d\tau\, \Phi_i \hat{H} \Phi_j \qquad (3.7.2)$$

---

* In certain cases symmetry requirements mean that more than one determinant must be used to describe a configuration, but the general idea is clear.

and

$$S_{ij} = \int d\tau\ \Phi_i\ \Phi_j$$

Re-arranging (3.7.1) we have

$$\sum_i S_{ij}\ \tilde{E} = \sum_i D_i\ D_j\ H_{ij} \qquad (j = 1,2,\ldots)$$

Forming $\frac{\partial E}{\partial D_j}$ for each value of j and equating each partial derivative to zero ensures a minimum in (3.7.1).

The resulting equations are best collected in the matrix form

$$(\mathbb{H} - \tilde{E}\mathbb{S})\mathbb{D} = 0 \tag{3.7.3}$$

or

$$\mathbb{H}\ \mathbb{D} = \tilde{E}\ \mathbb{S}\ \mathbb{D}$$

where $\mathbb{H}$ and $\mathbb{S}$ are the matrices whose elements are defined by (3.7.2) and $\mathbb{D}$ is a column matrix of the coefficients D_i. In particular, if we are able to choose a set of *orthogonal* Φ functions then $\mathbb{S} = \mathbb{1}$ and

$$\mathbb{H}\ \mathbb{D} = \tilde{E}\ \mathbb{D} \tag{3.7.4}$$

an obvious matrix analogue of the Schrödinger equation. This, and similar *matrix eigenvalue problems*, will play a prominent rôle in our development of orbital theories. The process of computing the eigenvectors ($\mathbb{D}$) and eigenvalues ($\tilde{E}$) of a matrix ($\mathbb{H}$) is usually referred to as the *diagonalisation* of the matrix. Thus the variational calculation associated with the Valence-Bond method reduces to a single matrix diagonalisation.

The application of the variation method to the optimisation of the orbitals of a single configuration is important enough to merit a separate chapter - Chapter 4.

3.8 ADDENDUM - THE FORMAL CONTENT OF CHAPTER 3

"Wittgenstein held mathematics to consist of equations which are dispensable in principle . . . . " *

Many chemists can find much more pressing and personal reasons than Wittgenstein's for regarding mathematics as dispensible. However, the ideas discussed in this chapter are the basis of all orbital theories and so a brief outline of the formal content of the results, derived earlier in a rather qualitative way, is in order.

Any set of orbitals which are the solutions of a one-electron Schrödinger equation form a complete set for the expansion of any continuous, normalisable function of three dimensions. That is, any function f(r) can be expanded as a linear combination of this set of orbitals:

$$f(r) = \sum_k b_k \mu_k(r) \qquad (3.8.1)$$

where the $\mu_k(r)$ are members of a complete set and the b_k are linear coefficients. If the orbitals form an orthogonal set

$$\int dr\ \mu_k(r)\mu_\ell(r) = \delta_{k\ell} \qquad (3.8.2)$$

then the formal relationship among the μ_k, b_k and f is

$$b_k = \int dr\ \mu_k(r) f(r) \qquad (3.8.3)$$

The summation in (3.8.1) is, for accurate expansion of arbitrary f, infinite. It is quite easy to show that the set of all possible products $\mu_k(r_1)\mu_\ell(r_2)$ form a complete set for the expansion of any continuous, normalisable function of six dimensions. The function to be expanded, $F_2(r_1,r_2)$ say, is written

$$F_2(r_1,r_2) = \sum_{k,\ell} B_{k\ell}\mu_k(r_1)\mu_\ell(r_2) \qquad (3.8.4)$$

---

* A.M. Quinton "Contemporary British Philosophy" in "A Critical History of Western Philosophy", ed. D.J. O'Connor. (MacMillan 1964)

If one variable is regarded as fixed at some value (R say) then the expansion becomes a function of only r_1:

$$F_2(r_1,R) = \sum_k C_k \mu_k(r_1)$$

where the coefficients C_k contain the information in the $B_{k\ell}$ for the particular value $r_2 = R$ chosen. This is an expression precisely analogous to (3.8.1) and so is valid. Considering the variable r_1 as fixed gives a similar valid expansion for the second variable r_2. Thus, provided that (3.8.1) holds, (3.8.4) is guaranteed. Proceeding in this way any function of n variables can be expanded in terms of the n-tuples which are the products of the $\mu_k(r)$:

$$F_n(r_1,r_2,\ldots r_n) = \sum_{i_1,i_2,\ldots i_n} B_{i_1 i_2 \ldots i_n} \mu_{i_1}(r_1)$$

$$\mu_{i_2}(r_2) \ldots \mu_{i_n}(r_n) \qquad (3.8.5)$$

again provided that the functions μ_k are a complete set in three dimensions. In the case of molecular wave functions the extra "spin" variable can be taken into account by noting that the two spin "functions" are a formal complete set in discrete spin "space". Thus any function of the n sets of four variables $(x_i, y_i, z_i, s_i) = x_i$ can be expanded in terms of products of the orbitals μ_k and a product of n spin factors $\zeta_k(s_i)$ where $\zeta_k(s_i)$ is $\alpha(s_i)$ or $\beta(s_i)$:

$$F'_n(x_1,x_2,\ldots x_n) = \sum_{i_1,i_2,\ldots,i_n} B'_{i_1 i_2 \ldots i_n}$$

$$\mu_{i_1}(r_1)\mu_{i_2}(r_2)\ldots\mu_{i_n}(r_n) \times \zeta_{i_1}(s_1)\zeta_{i_2}(s_2)\ldots\zeta_{i_n}(s_n)$$

$$(3.8.6)$$

where the prime has been added to the symbols F and B to denote the expansion of the space to include spin.

Any acceptable molecular wave function must satisfy the Pauli principle. Examination of the form of (3.8.6) shows that the antisymmetry requirement must be contained in the coefficients B'; each B' must be antisymmetric with respect to exchange of any pair of subscripts $i_1, i_2, \ldots i_n$ since the functions μ_k and ζ_i do not change when new variables are substituted. The coefficients B' are therefore redundant in the sense that sets of n! of them are related by antisymmetry in the subscripts. Associating a spin factor ζ_i with each spatial orbital μ_k and collecting the n! related terms as a determinant gives the spin-orbital expansion analogous to (3.6.1) where the D_k are the *essential* linear degrees of freedom:

$$F'_n(x_1, x_2, \ldots x_n) = \sum_k{}' D_k \, \Phi_k(x_1, x_2, \ldots x_n) \qquad (3.8.7)$$

where

$$\Phi_k(x_1, x_2, \ldots x_n) = \det\{\lambda_{i_1}(x_1)\lambda_{i_2}(x_2) \ldots \lambda_{i_n}(x_n)\}$$

The suffix k counts the number of distinct determinants obtainable from the occupation of an infinite complete set of spin-orbitals; $\lambda_i(x) = \mu_k(r)\zeta(s)$.

The formal similarity between (3.8.7) and (3.6.1) is encouraging - (3.8.7) guarantees that any n-electron molecular wave function can be expanded as a linear combination of determinants of spin-orbitals. The model partitioning of the Hamiltonian ensures that the functions μ_k are, in principle, a complete set, since they are solutions of a one-electron Schrödinger equation. *The orbital model for molecular electronic wave functions represents an application of the exact expansion (3.8.7) in which, for practical reasons, the expansion length (the number of determinants) is limited.*

SUGGESTIONS FOR FURTHER READING

"Quantum Theory of Many-Particle Systems I, II & III" (consecutive papers) by P.-O. Löwdin in Phys.Rev., 97 1474-1520 (1955) has a rather general discussion of orbital expansions and the determinantal method in density matrix language (note that Lowdin's density function gamma is our density function rho divided by p! - p is the number of particles).

4 ATOMIC ORBITALS

4.1 THE ONE-CONFIGURATION MODEL

The idea that each electron in a complex polyelectronic system can be assigned in some way to its "own" separate spatial distribution is a very attractive one and is retained throughout the *independent-electron model* of electronic structure. The term "independent-electron model" is rather unfortunate since any electron's distribution is, of course, very dependent on the spatial distribution of the other electrons. Since the term is very widely used we must perhaps think of the "independent" as meaning "independently assigned in the structure" and certainly not "independent of the motion of the other electrons". Within our orbital approximation scheme the independent-particle model is realised by using a wave function consisting of a single determinant of spin-orbitals; a single orbital configuration:

$$\Psi = \det\{\lambda_1(x_1)\lambda_2(x_2) \ldots \lambda_n(x_n)\} \qquad (4.1.1)$$

Here, each spin-orbital λ_i is occupied by just *one* electron; the n! terms in the determinant only differ by *which* electron is in λ_i. For the moment, we make the restriction that the electronic system shall be of "closed-shell" or "spin-paired" structure with each spatial orbital doubly occupied by two electrons of opposite spin. This restriction will be removed later to allow the treatment of "open-shell", paramagnetic systems.

In the following sections we shall use ϕ_i to represent a spatial *atomic* orbital in place of the general orbital notation μ_i. Thus we write

$$\lambda_1(x) = \phi_1(r)\alpha$$

$$\lambda_2(x) = \phi_1(r)\beta \qquad \text{etc.}$$

It is usual to use a contracted notation for spin-orbitals to avoid being swamped by typographical detail. The spin-orbital $\phi_i(r)\alpha(s)$ is written simply ϕ_i, and the combination $\phi_i(r)\beta(s)$ is written $\bar{\phi}_i$. A bar over a spatial orbital signifies that the occupying electron has β spin and no bar means α spin. Unless they are required for emphasis, the appearances of the explicit variables of the electrons - r_i and x_i - are also suppressed. There is little danger of confusion over the two possible meanings of ϕ_i - a spatial orbital and an α spin spin-orbital - since, in practice, the spin-orbital form usually only appears in a determinant. Using these contractions of notation equation (4.1.1) becomes

$$\Psi = \det\{\phi_1\ \bar{\phi}_1\ \phi_2\ \bar{\phi}_2 \cdots \phi_{n/2}\ \bar{\phi}_{n/2}\} \qquad (4.1.2)$$

The reasons for the rather severe compressions of notation become clear when the full form of (4.1.2) is written out for comparison:

$$\Psi = N\begin{vmatrix} \phi_1(r_1)\alpha & \phi_1(r_2)\alpha & \cdots & \phi_1(r_n)\alpha \\ \phi_1(r_1)\beta & \phi_1(r_2)\beta & \cdots & \phi_1(r_n)\beta \\ \cdots & \cdots & \cdots & \cdots \\ \cdots & \cdots & \cdots & \cdots \\ \phi_{n/2}(r_1)\alpha & \phi_{n/2}(r_2)\alpha & \cdots & \phi_{n/2}(r_n)\alpha \\ \phi_{n/2}(r_1)\beta & \phi_{n/2}(r_2)\beta & \cdots & \phi_{n/2}(r_n)\beta \end{vmatrix} \qquad (4.1.3)$$

(N is the normalising factor (3.3.5)).

Fortunately, the determinantal nature of the function (4.1.3) only has a rôle in determining the *form of the equations* satisfied by the component spatial functions ϕ_i. Thus, having decided to work within the one-configuration model for atoms, the computational problem is the evaluation of the atomic orbitals ϕ_i and we shall not make much further use of the determinantal notation in the derivations.

4.2 THE ROOTHAAN-HARTREE-FOCK METHOD FOR ATOMS

The atomic Hamiltonian is obtained as the special case of (2.2.1) when only one nucleus is present:

$$\hat{H} = \sum_{i=1}^{n} \hat{h}(i) + \sum_{i>j=1}^{n} \hat{g}(i,j)$$

where

$$\hat{h}(i) = -\tfrac{1}{2}\nabla^2(i) - \frac{Z}{r_i} \tag{4.2.1}$$

and

$$\hat{g}(i,j) = \frac{1}{r_{ij}}$$

(The notation Z_α and $r_{i\alpha}$ is redundant when only a single nucleus is involved and this nucleus is chosen to be the origin of coordinates).

The equations which the optimum orbitals ϕ_i must satisfy are obtained by substituting the one-configuration wave function (4.1.2) and the atomic Hamiltonian (4.2.1) into the variational expression. Minimisation of this variational expression with respect to the ϕ_i defines the optimum ϕ_i. Clearly, in order to carry through this program, we must assume a functional form for the atomic orbitals ϕ_i which:

(i) is physically realistic

and

(ii) contains adjustable parameters which can be optimised.

It is reasonable to suppose that the electronic structure of polyelectronic atoms can be based on the structure of the hydrogen atom: the forms of the atomic orbitals (universally abbreviated to AOs) are expected to be similar to the solutions of the Schrödinger equation for hydrogen-like "atoms". The hydrogen atom orbitals have the general form:

$$(\text{Polynomial in } r)\times(\text{Spherical Harmonic})\times\exp(-\zeta r) \qquad (4.2.2)$$

The central symmetry of atoms gives the Schrödinger equation a particularly simple form in spherical polar co-ordinates (r,θ,ϕ). The polynomials involved in (4.2.2) are functions of r in this system and the spherical harmonics are functions of θ and ϕ. The so-called *orbital exponent* ζ depends on the nuclear charge of the atom. Thus the one-electron orbitals are linear combinations of the following type:

$$(\text{Spherical Harmonic}) \times \sum_i b_i r^i \exp(-\zeta r)$$

where the coefficients b_i are *fixed by the solution of the one-electron Schrödinger equation*. If, therefore, we allow the coefficients b_i and the orbital exponents ζ to vary we have a plausible form for AOs of polyelectronic atoms which contains the freedom necessary for the application of the variation principle. That is, we shall use a linear combination of functions of the form:

$$r^{\nu} \exp(-\zeta r) \times (\text{Spherical Harmonic}) \qquad (4.2.3)$$

for each atomic orbital and optimise the overall wave function with respect to the linear coefficients and orbital exponents. In the first instance we consider the restricted optimisation obtained by allowing only the linear coefficients to vary.

Formally we write each atomic orbital ϕ_i as a linear combination of *basis functions* χ_i (say):

$$\phi_i = \sum_j c_j^{(i)} \chi_j \qquad (4.2.4)$$

where the linear coefficients $c_j^{(i)}$ can be collected into a column vector $\mathsf{C}^{(i)}$ and the basis functions defined as a row vector $\boldsymbol{\chi}$ giving

$$\phi_i = \boldsymbol{\chi}\,\mathsf{C}^{(i)} \tag{4.2.5}$$

$$\mathsf{C}^{(i)} = \begin{pmatrix} c_1^{(i)} \\ c_2^{(i)} \\ \cdot \\ \cdot \\ c_m^{(i)} \end{pmatrix} \tag{4.2.6}$$

$$\boldsymbol{\chi} = (\chi_1, \chi_2, \ldots \chi_m)$$

if we use m basis functions χ_i.

Since we require n/2 AOs (4.2.5) we can summarize the relation between the required n/2 AOs and the m basis functions as:

$$\boldsymbol{\varphi} = \boldsymbol{\chi}\,(\mathsf{C}^{(1)}, \mathsf{C}^{(2)}, \ldots \mathsf{C}^{(n/2)}) \tag{4.2.7}$$

here

$$\boldsymbol{\varphi} = (\phi_1, \phi_2, \ldots \phi_{n/2}).$$

Finally, defining

$$\mathsf{C} = (\mathsf{C}^{(1)}, \mathsf{C}^{(2)}, \ldots \mathsf{C}^{(n/2)})$$

we have

$$\boldsymbol{\varphi} = \boldsymbol{\chi}\,\mathsf{C}$$

as a summary of the n/2 equations (4.2.4); the matrix C contains the expansion coefficients which are to be optimised. The basis functions χ_i each have the general form (4.2.3) i.e.

$$\chi_i \sim r^{\nu} \exp(-\zeta r) \times \text{(Spherical Harmonic)}$$

Later, particularly in the study of molecules, we shall use different functional forms for the χ_i but the same general principle will apply: use of equation (4.2.4) for the orbitals and the optimisation of the matrix $\mathbf{C}$ by the variation method. Functions of this form (4.2.3) are called *S*later *T*ype *O*rbitals (STOs) after Slater who introduced them. From our point of view they are not orbitals but *basis functions*. However this name is likely to continue to be used - some authors use "Slater Functions" - and we shall not quibble.

In order to avoid over-complicating the derivation we assume that the basis functions $\boldsymbol{\chi}$ and the resultant atomic orbitals $\boldsymbol{\varphi}$ are orthogonal:

$$\int dr\ \chi_i(r)\chi_j(r) \ = \ \delta_{ij}$$

$$\int dr\ \phi_i(r)\phi_j(r) \ = \ \delta_{ij}$$

This restriction will be removed in Section 4.4.

The first step in the derivation is to obtain an explicit form for the approximate energy $\tilde{E}$ in terms of the functions $\boldsymbol{\varphi}$:

$$\tilde{E} \ = \ \frac{\int d\tau\ \Psi\ \hat{H}\ \Psi}{\int d\tau\ \Psi\ \Psi} \qquad (4.2.8)$$

Now Ψ contains n! orbital products and $\hat{H}$ consists of about n^2 terms therefore the expression for $\tilde{E}$ contains about $n^2(n!)^2$ terms! Fortunately, many of these terms have the same value - and many are zero since the functions $\boldsymbol{\varphi}$ are orthogonal. The final result of expanding (4.2.8) and collecting like terms is a special case of the Slater/Löwdin rules which we discuss in Chapter 5 in connection with Valence-Bond theory. In order to avoid duplication we simply quote the result:

$$\tilde{E} = 2\sum_{i=1}^{n/2} \{\int dr\phi_i(r)\hat{h}\phi_i(r)\} +$$

$$\sum_{i>j=1}^{n/2} \{2(\phi_i\phi_i,\phi_j\phi_j) - (\phi_i\phi_j,\phi_i\phi_j)\} \qquad (4.2.9)$$

In (4.2.9) we have introduced some standard notation for the energy integrals which arise in molecular calculations:

$$(\phi_i\phi_j,\phi_k\phi_\ell) = \int dr_1\int dr_2\phi_i(r_1)\phi_j(r_1)\hat{g}(1,2)\phi_k(r_2)\phi_\ell(r_2) \qquad (4.2.10)$$

Note that, in this definition $(\phi_i\phi_j,\phi_k\phi_\ell)$, the *order* of the two sets of orbitals is immaterial since from (4.2.10) the symmetric nature of $\hat{g}(1,2)$ ensures that

$$(\phi_i\phi_j,\phi_k\phi_\ell) = (\phi_k\phi_\ell,\phi_i\phi_j)$$

Since both the functions ϕ_i and ϕ_j depend on the coordinates of one electron and the functions ϕ_k and ϕ_ℓ depend on the other electron's variables we must have:

$$(\phi_i\phi_j,\phi_k\phi_\ell) = (\phi_j\phi_i,\phi_k\phi_\ell) = (\phi_i\phi_j,\phi_\ell\phi_k) = (\phi_j\phi_i,\phi_\ell\phi_k)$$

The order of ϕ_i and ϕ_j or ϕ_k and ϕ_ℓ does not matter in the definition of $(\phi_i\phi_j,\phi_k\phi_\ell)$. This notation for the integral (4.2.10) is often referred to as the "charge cloud" notation because of the interpretation of $\phi_i\phi_j$ and $\phi_k\phi_\ell$ as charge densities. The integral (4.2.10) is easily interpreted as the mean electrostatic repulsion energy between the density of electron 1 $(\phi_i\phi_j)$ and the density of electron 2 $(\phi_k\phi_\ell)$.

In order to show the explicit dependence of $\tilde{E}$ on the variables to be optimized we substitute from (4.2.4) into (4.2.9) obtaining

$$\tilde{E} = 2\sum_{i=1}^{n/2}\sum_{k,\ell=1}^{m} c_k^{(i)}c_\ell^{(i)}H_{k\ell} + \sum_{i=1}^{n/2}\sum_{k,\ell=1}^{m} c_k^{(i)}c_\ell^{(i)}G_{k\ell}$$

or, much more concisely, using the matrices $\mathsf{C}^{(i)}$ defined in (4.2.6)

$$\tilde{E} = 2\sum_i \mathsf{C}^{(i)\dagger}\,\mathsf{H}\,\mathsf{C}^{(i)} + \sum_i \mathsf{C}^{(i)\dagger}\,\mathsf{G}\,\mathsf{C}^{(i)} \qquad (4.2.11)$$

($\mathsf{C}^{(i)\dagger}$ is the transpose of $\mathsf{C}^{(i)}$).

The matrix H - elements $H_{k\ell}$ - is defined by the *basis function integrals*

$$H_{k\ell} = \int dr\ \chi_k(r)\hat{h}\chi_\ell(r) \qquad (4.2.12)$$

The contributions of the basis function integrals corresponding to (4.2.10) have been collected into the matrix G; the "total electron interaction matrix" of Roothaan:

$$G_k = \sum_{r,s=1}^{m} \{[2(k\ell,rs)-(kr,\ell s)]\sum_{i=1}^{n/2}(\mathsf{C}^{(i)}\mathsf{C}^{(i)\dagger})_{rs}\}$$

$$= \sum_{r,s=1}^{m} \{2(k\ell,rs)-(kr,\ell s)\}\,(\mathsf{C}\,\mathsf{C}^{\dagger})_{rs} \qquad (4.2.13)$$

The integrals involving the basis functions χ_i and the electron repulsion operator $\hat{g}(1,2)$ have been abbreviated to $(k\ell,rs)$ where

$$(k\ell,rs) = \int dr_1\int dr_2\chi_k(r_1)\chi_\ell(r_1)\hat{g}(1,2)\chi_r(r_2)\chi_s(r_2) \qquad (4.2.14)$$

The matrix H is usually referred to as the *one-electron Hamiltonian matrix* and the basis function integrals $(k\ell,rs)$ as the *electron repulsion integrals* or, more commonly, the *two-electron integrals*.

Returning to the derivation, we must optimise the coefficients $\mathsf{C}^{(i)}$ by minimising (4.2.11) with respect to variations in the coefficients consistent with the normalisation and orthogonality of the functions ϕ_i. Since we have assumed that the functions χ_i are orthogonal, it is easy to show that

$$C^{(i)\dagger}\,C^{(j)} = \delta_{ij} \tag{4.2.15}$$

summarizes the orthogonality and normalization conditions among the ϕ_i. If we vary the coefficients defining the ith atomic orbital by $\delta C^{(i)}$ then the associated variation in $\tilde{E}$ (dropping quadratic and higher terms in $\delta C^{(i)}$) is

$$\begin{aligned}\delta\tilde{E}^{(i)} &= 2\delta C^{(i)\dagger}\,H\,C^{(i)} + 2C^{(i)}\,H\,\delta\,C^{(i)}\\ &+ \delta C^{(i)\dagger}\,G\,C^{(i)} + C^{(i)\dagger}\,\delta\,G\,C^{(i)} + C^{(i)\dagger}\,G\,\delta\,C^{(i)}\end{aligned} \tag{4.2.16}$$

Now

$$\delta C^{(i)\dagger}\,H\,C^{(i)} = C^{(i)\dagger}\,H\,\delta\,C^{(i)}$$

and

$$\delta C^{(i)\dagger}\,G\,C^{(i)} = C^{(i)\dagger}\,G\,\delta\,C^{(i)}$$

Also

$$(\delta\,G)_{k\ell} = 2\sum_{r,s=1}^{m}\{[2(k\ell,rs)-(kr,\ell s)]\,(\delta C^{(i)}C^{(i)\dagger})_{rs}\}$$

Hence

$$\delta C^{(i)\dagger}\,G\,C^{(i)} = C^{(i)\dagger}\,\delta\,G\,C^{(i)}$$

Collecting these results, we can simplify (4.2.16) to

$$\delta\tilde{E}^{(i)} = 4\,\delta\,C^{(i)\dagger}\,H^{F}\,C^{(i)} \tag{4.2.17}$$

where

$$H^{F} = H + G$$

For a minimum in $\tilde{E}$ each $\delta\tilde{E}^{(i)}$ must vanish separately, consistent with variations in (4.2.15)*

---

* Strictly we should use $\delta C^{(i)\dagger}C^{(j)}=0$ in place of (4.2.18) but *orthogonality* of ϕ_i is automatic and does not have to be imposed.

i.e. $$C^{(i)\dagger} C^{(i)} = 0 \tag{4.2.18}$$

Combining (4.2.17) and (4.2.18) gives

$$\delta C^{(i)\dagger} (H^F C^{(i)} - \varepsilon_i H^F C^{(i)}) = 0 \tag{4.2.19}$$

where ε_i is a Lagrange multiplier.

If (4.2.19) is to hold for *arbitrary* variations $\delta C^{(i)}$ then the contents of the bracket must vanish independently of $\delta C^{(i)}$:

$$H^F C^{(i)} = \varepsilon_i C^{(i)} \qquad (i = 1,2,\ldots n/2) \tag{4.2.20}$$

The whole set of n/2 equations can be combined by using (4.2.7) and the definition of C:

$$H^F C = C \boldsymbol{\epsilon} \tag{4.2.21}$$

where $\boldsymbol{\epsilon}$ is a diagonal matrix of the ε_i.

The original problem, the minimization of (4.2.9), is equivalent to a matrix eigenvalue problem - the computation of the eigenvalues and eigenvectors of the matrix H^F. In fact, the notation we have used disguises the fact that the problem is not quite so simple as a single matrix diagonalisation since H^F contains C through the matrix G. Any method of solution of (4.2.21) must therefore be *iterative;* H^F and C must be found which *both* satisfy (4.2.21) *self consistently*. The iterative nature of the solution of (4.2.21) gives the resulting atomic orbitals their familiar name: *S*elf *C*onsistent *F*ield *A*tomic *O*rbitals (SCFAOs). Equation (4.2.21) is a matrix form of the Roothaan-Hartree-Fock equations (the RHF equations) and H^F is the Roothaan-Hartree-Fock matrix. In the next two sections we will discuss the physical interpretation of the RHF equations and sketch the practical steps involved in their solution.

4.3 THE INTERPRETATION OF THE RHF EQUATION

Equation (4.2.21) is simply a convenient way of collecting the n/2 equations (4.2.20) and so the interpretation of the terms in (4.2.20) will be discussed:

$$H^F C^{(i)} = \varepsilon_i C^{(i)} \tag{4.3.1}$$

where

$$H^F = H + G \tag{4.3.2}$$

First the column vector $C^{(i)}$: this is simply the collection of optimum coefficients of the basis orbitals χ defining the ith atomic orbital ϕ_i:

$$\phi_i = \chi C^{(i)}$$

Postponing for the moment the iterative solution of (4.3.1) and regarding the equations as solved, the matrix H^F must be the matrix representation of some "Hamiltonian" operator in the basis χ. The definition (4.3.2) gives the physical interpretation; clearly H is the one-electron Hamiltonian, the elements

$$H_{ij} = \int dr\ \chi_i(r)\hat{h}\chi_j(r) \tag{4.3.3}$$

measuring the mean kinetic energy and nuclear attraction energy of an electron with spatial distribution $\chi_i\chi_j$. The matrix G, which contains all reference to the electron-repulsion integrals, must represent the mean repulsion energy of an electron with density $\chi_i\chi_j$. Re-writing equation (4.2.13) in an expanded form we have

$$G_{k\ell} = 2\sum_{r,s=1}^{m}\sum_{i=1}^{n/2}(C^{(i)}C^{(i)\dagger})_{rs}\int dr_1\int dr_2\chi_k(r_1)\chi_\ell(r_1)\frac{1}{r_{12}}\chi_r(r_2)\chi_s(r_2)$$

$$-\sum_{r,s=1}^{m}\sum_{i=1}^{n/2}(C^{(i)}C^{(i)\dagger})_{rs}\int dr_1\int dr_2\chi_k(r_1)\chi_r(r_1)\frac{1}{r_{12}}\chi_\ell(r_2)\chi_s(r_2) \tag{4.3.4}$$

The first line of this rather lengthy expression can be re-arranged to

$$2\int dr_1 \left(\int dr_2 \sum_{r,s=1}^{m} \sum_{i=1}^{n/2} (C^{(i)}C^{(i)\dagger})_{rs} \chi_r(r_2)\chi_s(r_2)\frac{1}{r_{12}} \right) \chi_k(r_1)\chi_\ell(r_1) \tag{4.3.5}$$

This form of the expression shows that it is the mean repulsion energy of the charge distribution $\chi_k\chi_\ell$ with the charge density in curly brackets - the sum of all the possible products $\chi_r\chi_s$ multiplied by the weights with which they occur in the total electron density. *It is the mean repulsion energy between $\chi_k\chi_\ell$ and the rest of the electronic system.* The second term in (4.3.4) - often called the *exchange term* because of the way it arises in the derivation of the energy expression - corrects for the self-repulsion term included in (4.3.5) and also includes the effect of the Pauli principle on the orbital model. The sum of the two terms is the mean potential which an electron having distribution $\chi_k\chi_\ell$ experiences due to all the other electrons. Adding this potential to the one-electron Hamiltonian gives the total potential "seen" by an electron with distribution $\chi_k\chi_\ell$.

The method of solution of (4.3.1) based on making a guess at the $C^{(i)}$ and iteratively improving them has, therefore, some physical parallel in seeking the stable arrangement of the electron distribution based on an assumed non-equilibrium starting point.

The eigenvalues ε_i of (4.3.1) have the dimensions of energy and must in some way be a measure of the energy of an electron associated with orbital ϕ_i (coefficients $C^{(i)}$). The precise meaning of these *orbital energies* is easily found by analysing the effect on the total energy of removing an electron from ϕ_i. The total energy of the positive ion having n-1 electrons is given by

$$E^+ = 2\sum_{i\neq j=1}^{n/2} C^{(j)\dagger} H\, C^{(j)} + \sum_{i\neq j=1}^{n/2} C^{(j)\dagger} G\, C^{(j)}$$

$$+\, C^{(i)\dagger} H\, C^{(i)} + C^{(i)\dagger} G\, C^{(i)} \tag{4.3.6}$$

Therefore the energy difference between the atom and the ion is

$$E - E^{+} = C^{(i)\dagger} H^{F} C^{(i)} = \varepsilon_i \qquad (4.3.7)$$

using (4.3.2) and (4.2.11).

The orbital energy ε_i is therefore minus the ionisation potential of an electron in ϕ_i. This result is known as Koopmans' theorem. The ε_i are not quite the best ionisation potentials which can be computed since, in writing (4.3.6), we have assumed that the coefficients defining an orbital ($C^{(i)}$) *do not change* on ionisation. Since each set of orbital coefficients depend on all the others we should really perform a separate calculation on the positive ion to get the best results for E^{+}.

4.4 THE USE OF NON-ORTHOGONAL BASIS FUNCTIONS

In deriving equation (4.2.21) we made use of the fact that the functions χ had been assumed to be orthogonal - equation (4.2.15). The usual situation is that the basis functions are not orthogonal and generate a so-called *overlap matrix* S with elements

$$S_{ij} = S_{ji} = \int dr \chi_i(r) \chi_j(r) \qquad (4.4.1)$$

In this case the overlap integrals between orbital ϕ_i and orbital ϕ_j are given by

$$C^{(i)\dagger} S \, C^{(j)} \quad ,$$

the orthogonality condition (4.2.15) is replaced by

$$C^{(i)\dagger} S \, C^{(j)} = \delta_{ij} \qquad (4.4.2)$$

and the effect of a variation in orbital ϕ_i maintaining normalisation is

$$\delta C^{(i)\dagger} S \, C^{(i)} = 0 \qquad (4.4.3)$$

which replaces (4.2.18). Using this new normalisation condition the RHF equation becomes

$$H^F C = S C \epsilon \qquad (4.4.4)$$

where the overlap matrix plays a vital rôle in the definition of the atomic orbitals. The solution of equation (4.4.4) will be discussed as an application of theorems derived in Chapter 9.

4.5 SUMMARY

The unavoidably mathematical nature of the present chapter may have masked the essential steps in the solution of the independent-electron model for atomic systems. The following is a bald outline of the previous sections.

(i) The model wavefunction is a single determinant of (doubly-occupied spatial) orbitals

$$\Psi = \det\{\phi_1\bar{\phi}_1\phi_2\bar{\phi}_2 \ldots \phi_{n/2}\bar{\phi}_{n/2}\} \qquad (4.1.2)$$

(ii) The atomic orbitals $\boldsymbol{\varphi}$ are expanded in terms of a set of basis functions $\boldsymbol{\chi}$ (chosen on physical grounds)

$$\boldsymbol{\varphi} = \boldsymbol{\chi} C \qquad (4.2.7)$$

(iii) The expansion coefficients in the matrix C are optimised using the variation principle. This leads to the definition of the Roothaan-Hartree-Fock equations for the matrix C:

$$H^F C = C \epsilon \qquad (4.2.21)$$

(or $H^F C = S C \epsilon$ (4.4.4))

(iv) The orbital energies ε_i are the ionisation potentials of the orbitals (apart from sign).

In practice, the spherical symmetry of atoms enables special techniques to be used to solve the RHF equation but we have given the full derivation in anticipation of the Molecular Orbital method for molecules. It is also possible to compute atomic orbitals directly from the differential equations by numerical methods. These numerical atomic orbitals do not have direct application in molecular calculations and so we have not discussed them. The atomic orbitals which are solutions of (4.2.21) are the basic units of any "valence" theory of molecular electron distribution.

SUGGESTIONS FOR FURTHER READING

"Recent Developments in Molecular Orbital Theory" by C.C.J. Roothaan in Rev.Mod.Phys., 23 69 (1951) contains derivation of the "RHF Equation" in the context of Molecular Orbital Theory.

Solutions of the RHF equations for atoms He through Krypton are described and tabulated in "Ab initio Computations in Atoms and Molecules" IBM Journal Research and Development 9 2 (1965) by E. Clementi and supplement (available on request from IBM Research Laboratory, San Jose, California 95114, U.S.A.)

The theory and practice of the computation of numerical atomic orbitals is given in "Calculation of Atomic Structures" by D.R. Hartree (Wiley)

Numerical orbitals and energy quantities for all the atoms from Hydrogen to Lawrencium are given in "Atomic Structure Calculations" by J.B. Mann, Los Alamos Technical Report LA3690 and 3691 (available from N.B.S., Springfield, Virginia 22151, $6.00).

5 THE MOLECULAR ORBITAL AND VALENCE BOND METHODS

5.1 SURVEY OF THE MOLECULAR ORBITAL AND VALENCE BOND METHODS

In order to satisfy our conditions that valence theory should be a theory of *changes* in electron distribution on bonding, and to remain within an orbital theory, the atomic orbitals will be the basis of a description of molecular electronic structure. Ideally, a theory of valence should use accurate atomic wave functions as units in molecular wave functions, but the intuitive and historical appeal of the independent electron model is so great that AO's are invariably used. One speaks, at a qualitative level, of "a 2p electron in ethylene" or "the 3d electrons in nickel tetracarbonyl" - almost unconsciously using the one-configuration model. The various "weights" with which the AO's of the constituent atoms of a molecule occur in the molecular wave function gives a direct *valence* interpretation.

Having decided to use AO's, we now turn to the possible ways of using them in forming approximate molecular wave functions of the form (3.6.1)

$$\Psi = \sum_i D_i \Phi_i \qquad (5.1.1)$$

where the Φ_i functions are determinants of orbitals. We have previously noted in Section 3.6 that the optimisation of *both* the orbitals in Φ_i *and* the linear coefficients D_i is out of reach for molecules of chemical interest and we now consider two alternatives

(i) The molecular analogue of the atomic RHF method: we take a single term of (5.1.1) and optimise the orbitals of the determinant. This is the Molecular Orbital (MO) method and we use the symbol ψ_i to represent a typical spatial MO in place of the general orbital notation μ_i (cf. ϕ_i for AO's).

(ii) We retain the multi-configuration form of (5.1.1) and optimise the linear coefficients D_i for a *fixed* set of spatial orbitals. As we have noted previously, this is one mathematical formulation of the familiar Valence Bond (VB) method.

In principle, we would like to use the most suitable of these two methods for a particular application; in practice the method which is most easily adapted to computer implementation will be used most widely.

Some light can be thrown on the physical nature of the two approximation schemes by considering them as being developed from two possible ways of partitioning the molecular Hamiltonian operator (2.2.1).

If we divide the terms in the Hamiltonian into the "one-electron" and "two-electron" sets as follows:

$$\hat{H} = \left[\sum_{i=1}^{n} (-\tfrac{1}{2}\nabla^2(i) - \sum_{\alpha=1}^{N} \frac{Z_\alpha}{r_{i\alpha}}) \right] + \left[\sum_{i>j=1}^{n} \frac{1}{r_{ij}} \right] \tag{5.1.2}$$

and neglect the second, electron repulsion, term the resulting equation is one of the type discussed in Chapter 3 and one solution is a determinant of spin-orbitals. The spatial parts of these orbitals (ψ_i) are solutions of

$$\left(-\tfrac{1}{2}\nabla^2 - \sum_{\alpha=1}^{N} \frac{Z_\alpha}{r_{i\alpha}} \right) \psi_i = \varepsilon_i \psi_i \tag{5.1.3}$$

The potential energy term in this one-electron Schrödinger equation ensures that the orbitals ψ_i are *many-centre*; that is, they have significant values over the whole molecule and are not, for example, centred mainly around a single atom. Thus, if we choose a suitable many-centre form for the spatial molecular orbitals ψ_i which contains adjustable parameters, and form a single determinant of spin orbitals formed from the ψ_i, we have a trial function with which to approximate the solution of the Schrödinger equation associated with (5.1.2). The inclusion of the electron-repulsion terms in $\hat{H}$ means that the optimum one configuration orbitals will not be solutions of (5.1.3) but they will still have the characteristic feature of being spread over the whole molecule ("delocalised" or "canonical" molecular orbitals). The delocalised form of the molecular orbitals has obvious advantages in describing molecular properties or processes involving the over-all electronic structure of the molecule - spectroscopic transitions, ionisation, etc.

The full Hamiltonian (5.1.2) can also be partitioned in a more "chemical" way; into essentially atomic and interatomic terms:

$$\hat{H} = \sum_{\alpha=1}^{N} \hat{H}_\alpha + \sum_{\alpha>\beta=1}^{N} \hat{H}_{\alpha\beta}$$

where

$$\hat{H}_\alpha = \sum_{i\varepsilon\alpha} \left(-\tfrac{1}{2}\nabla^2(i) - \frac{Z_\alpha}{r_{i\alpha}}\right) + \sum_{i>j\varepsilon\alpha} \frac{1}{r_{ij}}$$

is the Hamiltonian for atom α and

$$\hat{H}_{\alpha\beta} = -\sum_{i\varepsilon\alpha} \frac{Z_\beta}{r_{i\alpha}} - \sum_{j\varepsilon\beta} \frac{Z_\alpha}{r_{j\beta}} + \sum_{i\varepsilon\alpha} \sum_{j\varepsilon\beta} \frac{1}{r_{ij}}$$

is the term representing the interaction between the electrons of atom α and those of atom β with each other and each others' nucleus. The notation $i\varepsilon\alpha$ means that the summation is to be carried out over the variables of those electrons associated with nucleus α. In this case neglect of the second term leads to a set of different atomic Hamiltonians $\hat{H}_\alpha$ and associated

Schrödinger equations. If we solve each of the atomic Schrödinger equations by the RHF method we obtain a set of atomic orbitals for each atom. This suggests that an approximate solution of the molecular Schrödinger equation would be given by a *determinant of atomic orbitals*. Since the atomic orbitals are fixed by the nature of the atoms in the molecule, the required degrees of freedom needed to optimise the molecular wave function variationally can be given by using a *linear combination* of determinants of atomic orbitals with adjustable coefficients - the D_i of (5.1.1). This approach uses atomic orbitals throughout and so corresponds to a more "localised" description of chemical bonding in terms of electron configurations or "chemical structures" as they are called in the VB context. The VB method has extensive qualitative applications, particularly in organic chemistry. If these rather intuitive arguments seem flimsy justification for a whole program of molecular calculations we can again fall back on the formal arguments of section 3.8.

5.2 THE MOLECULAR-ORBITAL (MO) METHOD

The aim of the MO method is to find the best possible one-configuration (single-determinant) approximate solution of

$$\hat{H}\,\Psi = E\,\Psi$$

where $\hat{H}$ is the full non-relativistic molecular Hamiltonian, and Ψ is a determinant of spin-orbitals whose spatial components are the MO's ψ_i. The simplest and most commonly occurring case is the closed-shell ground-state function,

$$\Psi = \det\{\psi_1\bar{\psi}_1\psi_2\bar{\psi}_2 \ldots \psi_{n/2}\bar{\psi}_{n/2}\} \qquad (5.2.1)$$

in the notation of (4.1.2). This closed-shell case involves the bulk of organic and non-transition-metal inorganic chemistry.

The computational method is to assume a physically plausible form for the functions ψ_i, which contains adjustable parameters, and optimise these parameters using the variation principle. The mathematics is therefore identical to the atomic RHF method solved in Chapter 4 with one important difference. Atoms are all the same "shape" and therefore a general functional form can be used for the basis functions used to expand the atomic orbitals - the form given in 4.2 is the most common. Molecules have different shapes and symmetries and so any basis functions used to expand the molecular orbitals can not be so general. It is easy to guess a limiting form for the MO's in the region of each atom: the molecular orbitals must pass over into atomic orbitals close to each nucleus. What better form for the approximate MO's, on both mathematical and chemical grounds, than a Linear Combination of Atomic Orbitals (LCAO) - a linear combination of the orbitals of the constituent atoms of the molecule? This choice of approximate molecular orbitals is a further approximation over and above the MO method but it does lead to a convenient *valence* interpretation of the molecular wave function.

We therefore write each MO ψ_i in terms of the AO's ϕ_j

$$\psi_i = \sum_j T_{ji}\phi_j \qquad (i=1,2,\ldots n/2) \qquad (5.2.1)$$

or, collecting the MO's as a row vector $\boldsymbol{\psi}$,

$$\boldsymbol{\psi} = \boldsymbol{\varphi}\,\mathsf{T}$$

The elements T_{ji} of the matrix T are then optimised using the variation principle, and the optimum matrix defines the structure of the molecular orbitals in terms of the atomic orbitals.

The atomic orbitals $\boldsymbol{\varphi}$ are expanded in terms of a set of basis functions $\boldsymbol{\chi}$, and the MO's in turn expand as linear combinations of the AO's.

$$\boldsymbol{\varphi} = \boldsymbol{\chi}\,\mathsf{C}$$

$$\boldsymbol{\psi} = \boldsymbol{\varphi}\,\mathsf{T} = \boldsymbol{\chi}(\mathsf{C}\,\mathsf{T}) \qquad (5.2.2)$$

Equation (5.2.2) suggests an alternative method of computing molecular orbitals: the MO's could be expanded *directly in terms of the sets of basis functions centred on each atom:*

$$\psi = \chi \mathbb{M} \tag{5.2.3}$$

where the matrix $\mathbb{M}$ contains the coefficients relating each MO to the atomic basis functions. A valence interpretation of this form of MO method is possible by comparing the coefficients of the two matrices $\mathbb{C}$ and $\mathbb{M}$ which give the changes in electron distribution on going from the atoms to the molecule. Both approaches are equally valid, the latter method - use of (5.2.3) - having the slight advantage of greater flexibility since the set χ has more members than the set φ. The relation among the functions χ, φ and ψ can be represented diagramatically as:

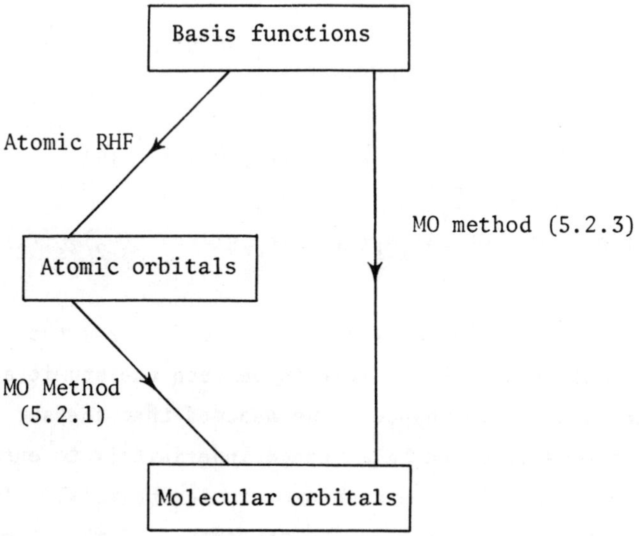

Specifically, we assume the LCAO MO method - the use of (5.2.1). In this case the chemical information is available in the most direct way - the elements of $\mathbb{T}$ show how the AO's are combined to form MO's. (Current usage is LCAO MO for *both* (5.2.1) and

and (5.2.3) even though only (5.2.1) uses atomic orbitals).

The technique of deriving the equations which the optimum T matrix must satisfy is obviously completely analogous to the calculation of the atomic matrix C and the results - the *molecular* RHF equation - can be obtained simply by a change of notation.

$$\mathsf{H}^F \mathsf{T} = \mathsf{T}\boldsymbol{\epsilon} \tag{5.2.4}$$

$$(\text{or} \quad \mathsf{H}^F \mathsf{T} = \mathsf{S}\mathsf{T}\boldsymbol{\epsilon}) \tag{5.2.5}$$

where the elements of H^F, T, S and $\boldsymbol{\epsilon}$ are expressed in terms of the AO's $\boldsymbol{\varphi}$ in place of the basis functions

$$S_{ij} = \int dr\phi_i(r)\phi_j(r) \tag{5.2.6}$$

$$\mathsf{H}^F = \mathsf{H} + \mathsf{G} \tag{5.2.7}$$

$$H_{ij} = \int dr\phi_i(r)\hat{h}\phi_j(r) \tag{5.2.8}$$

$$G_{ij} = \sum_{r,s} (\mathsf{T}\,\mathsf{T}^\dagger)_{rs}[2(ij,rs)-(ir,js)] \tag{5.2.9}$$

$$(ij,rs) = \int dr_1\int dr_2\phi_i(r_1)\phi_j(r_1)\hat{g}(1,2)\phi_r(r_2)\phi_s(r_2)$$

Although (5.2.4) is formally analogous to (4.2.20) there is an important difference of application between the atomic and molecular cases. In Chapter 4 we assumed that the set of basis functions $\boldsymbol{\chi}$ could be extended indefinitely to ensure that the functions $\boldsymbol{\varphi}$ were accurate atomic orbitals. In molecules, the set $\boldsymbol{\varphi}$ is fixed by the nature of the atoms in the molecule and so our MO's $\boldsymbol{\psi}$ will not be the best possible molecular orbitals - they are limited by the capabilities of the AO functions. If they are solutions of (5.2.4), however, they will be the best possible MO's *consistent with the limited set of AO's available* and, for convenience, we shall still refer to them as "solutions of the molecular RHF equation". The set $\boldsymbol{\varphi}$ can, of course, be extended (in principle indefinitely) by the addition of "excited orbitals" of the atoms but in this case the chemical interpretation becomes more difficult. For many purposes it is vital to involve the *low-lying* excited orbitals of the component atoms of a molecule - we shall do this in the BeH_2 example later.

The chemical information in the matrix T is made much more accessible by the definition of a new matrix R which defines the *weights* with which the various orbital products ("electron densities") appear in the total electron density. If we ask for the total "occupation" of a typical atomic orbital ϕ_i in the molecule the answer can be given in terms of the coefficient with which ϕ_i^2 appears in the total electron density P(r) of (2.6.5). Now a given AO will, in general, occur in all the MO's with various coefficients (elements of T). The sum of the squares of these coefficients of ϕ_i in all the ψ_j gives the total "occupation number" of AO ϕ_i. This is just the sum of the squares of the j'th *row* of T doubled; since each spatial MO is doubly occupied:

$$2\left(\sum_{j=1}^{n/2} T_{ij}^2\right)\phi_i^2$$

is the contribution to the total electron density from electrons in ϕ_i. In chemistry we are also interested in the electron density in the internuclear regions - the bonding densities - and so a similar interpretation can be given to the coefficient of $\phi_i\phi_j$ in the total electron density. The expression

$$2\sum_{k=1}^{n/2} T_{ik}T_{jk}\phi_i\phi_j$$

is the contribution of the electrons occupying the region of overlap of ϕ_i and ϕ_j and the numerical coefficient can be interpreted as the population of this overlap region. Both these definitions are neatly summarised by defining

$$\mathsf{R} = \mathsf{T}\mathsf{T}^\dagger \tag{5.2.10}$$

The occupation number of any orbital product $\phi_i\phi_j$ is given by $2R_{ij}$. As we have seen, the diagonal elements of R represent the number of electrons in each atomic orbital; they are called the orbital *charges*. The off-diagonal elements, summarising the population of the bonding regions, are known as the *bond orders*. R, or sometimes $2\mathsf{R}$, is known as the *charge and bond order matrix*.

Our interpretation of the elements of R as defining the various contributions to the electron density suggests that R must be related to the density function P(r) defined by (2.6.5). Using our one-determinant approximate function Ψ in (2.6.5) and carrying through the integrations is a straightforward but tedious exercise - the result is:

$$P(r) = 2\boldsymbol{\varphi}(r)\mathsf{R}\boldsymbol{\varphi}^\dagger(r) \tag{5.2.11}$$

(where we have added the explicit dependence of the AO's on r for emphasis).

Written out in full (5.2.11) becomes

$$P(r) = 2\sum_{i,j} R_{ij}\phi_i(r)\phi_j(r)$$

The matrix R is a finite matrix representation of the total electron density *in terms of the AO functions* ϕ_i. The *spatial dependence* of the electron density has been absorbed into the AO products $\phi_i\phi_j$ leaving the population *numbers* R_{ij} summarising the "large scale" electron distribution. We shall return to the interpretative value of R in Chapter 10.

Examination of the definition of the molecular G matrix in (5.2.9) shows that the functional dependence of G is on R rather than T. In fact, substituting from (5.2.10) into (5.2.9), we have

$$G_{ij} = \sum_{r,s} R_{rs}[2(ij,rs) - (ir,js)] \qquad (5.2.12)$$

Because of the simplicity of (5.2.12) compared to (5.2.9) (or (4.2.12)) it is usual to think of R as the independent variable defining G and to emphasise this functional dependence by writing the electron interaction matrix as $\mathsf{G}(\mathsf{R})$. The expression for the total electronic energy of a molecular-orbital wave function is analogous to the atomic expression (4.2.10). Using the definition of R it is easy to show that the electronic energy is given by

$$E = 2\mathrm{tr}\,\mathsf{H}\,\mathsf{R} + \mathrm{tr}\,\mathsf{G}(\mathsf{R})\mathsf{R} \qquad (5.2.13)$$

[The notation "tr" indicates the operation of summing the diagonal elements - the trace - of the following matrix, that is,

$$\mathrm{tr}\mathsf{A} = \sum_i A_{ii}$$

The trace of a product of two matrices A and B is, of course,

$$\mathrm{tr}\,\mathsf{A}\,\mathsf{B} = \sum_{i,j} A_{ij}B_{ji}\ ].$$

In the context of (5.2.13) it is possible to regard R, not T, as the independent variable to be optimised in minimising E. It is a useful exercise to show that the analogue of (5.2.4) is

$$\mathsf{H}^F\mathsf{R} - \mathsf{R}\mathsf{H}^F = 0 \qquad (5.2.14)$$

and that the orthogonality and normalisation constraints,

$$\mathsf{T}^\dagger\mathsf{T} = \mathsf{1}$$

are equivalent to

$$\mathsf{R}^2 = \mathsf{R} \qquad (5.2.15)$$

5.3 THE VALENCE BOND (VB) METHOD

The literature on the theory of the VB method is enormous and has its origin at the very beginning of quantum chemistry. From our point of view the various formulations and re-formulations of the VB problem in the language of a variety of mathematical structures is not so important as the fact that they all reduce to the same *computational problem*. However, since we have asserted several times that equation (5.1.1) *is* the VB method, the connection between this equation and the resonance diagrams of the chemist perhaps merits a short digression.

The Heitler-London method for a homopolar two-electron bond "formed" by the combination of two atomic orbitals is the theoretical basis of the single lines used in chemical primary structure diagrams. If the two atoms are A and B, and the two AO's ϕ_A and ϕ_B:

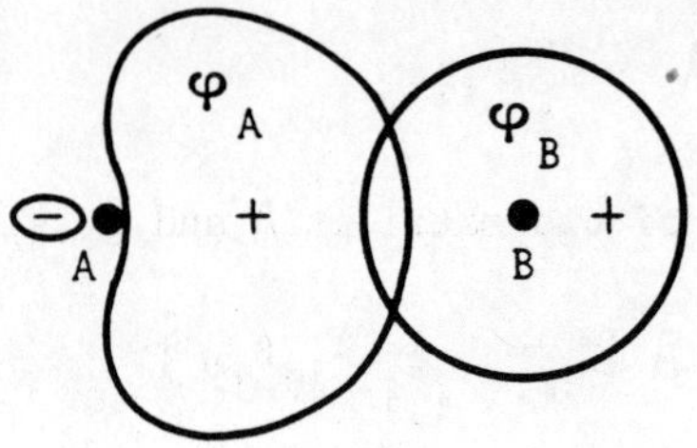

then the Heitler-London wave function for the lowest state of the whole system is

$$\Psi_{HL} = N_1[\phi_A(r_1)\phi_B(r_2) + \phi_A(r_2)\phi_B(r_1)]$$

$$\times [\alpha(s_1)\beta(s_2) - \beta(s_1)\alpha(s_2)] \qquad (5.3.1)$$

(N_1 is a normalising factor).

The chemists' notation for such a function is A—B. If this description is found to be inadequate, particularly likely if the electronegativities of A and B are different, then the polar structures

$$A^+ \quad B^-$$

or

$$A^- \quad B^+$$

would be postulated by the chemist. Corresponding wave functions formed from ϕ_A and ϕ_B are

$$\Psi_{+-} = N_2\phi_B(r_1)\phi_B(r_2)[\alpha(s_1)\beta(s_2) - \beta(s_1)\alpha(s_2)] \quad (5.3.2)$$

and

$$\Psi_{-+} = N_3\phi_A(r_1)\phi_A(r_2)[\alpha(s_1)\beta(s_2) - \beta(s_1)\alpha(s_2)] \quad (5.3.3)$$

The over-all electronic structure can then be written in the two equivalent forms:

$$A - B \leftrightarrow A^+ B^- \leftrightarrow A^- B^+$$

and (5.3.4)

$$\Psi = a\Psi_{HL} + b\Psi_{+-} + c\Psi_{-+}$$

where the numerical coefficients a, b and c are to be optimised.

Using the same two AO's and the formulation (5.1.1) we obtain

$$\Psi = D_1 \det\{\phi_A\bar{\phi}_A\} + D_2 \det\{\phi_B\bar{\phi}_B\}$$

$$+ D_3 \det\{\phi_A\bar{\phi}_B\} + D_4 \det\{\bar{\phi}_A\phi_B\} \quad (5.3.5)$$

Judicious manipulation of the expanded forms of (5.3.5) and (5.3.4) show them to be the same, confirming our use of the term "VB" for (5.1.1): the expansion of the molecular wave function in terms of determinants of AO's.

In the more general case of a polyelectronic molecule for which there are many possible bonding schemes (structures) - the assignment of many electrons among many AO's - there is obviously a problem of "choice of structures" or "choice of determinants". Only if all possible structures and determinants are included are the analogues of (5.3.4) and (5.3.5) identical. The various ways of formulating the VB method differ in the language they use to describe the choice of structure. The determinantal method is a flexible and convenient way of expanding a VB wave function as all formulations are easily "translated" into determinantal "language" by a procedure analogous to the comparison of (5.3.4) and (5.3.5). For example, a structure consisting of n orbitals describing a system of n/2 homopolar bonds each with a Heitler-London type wave function is:

$$\Psi = A \prod_{i=1,3..}^{n/2} [\phi_i(r_i)\phi_{i+1}(r_{i+1}) + \phi_{i+1}(r_i)\phi_i(r_{i+1})][\alpha\beta-\beta\alpha] \quad (5.3.6)$$

This function, when expanded, is equivalent to a fixed linear combination of $2^{n/2}$ determinants. Similar results hold for polar structures of the type (5.3.2) and so a consideration of the wave function defined by (5.1.1) is adequate.

In summary, the steps in the VB method are:

i) Take the set of all AO's of the atoms in the molecule together with the spin factors α, β and form a set of spin-orbitals.

ii) Form determinants from selection of the spin-orbitals which reflect the likely electronic structure of the molecule (in terms of bonds, lone pairs, inner shells).

iii) Form a linear combination of these determinants (5.1.1) and optimise the numerical coefficients.

We have seen in Section 3.7 that the last step involves the linear variation method - the solution of

$$\mathsf{H}\,\mathsf{D} = \mathsf{S}\,\mathsf{D}\,E$$

where

$$H_{ij} = \int d\tau\, \Phi_i\, \hat{H}\, \Phi_j$$

and (5.3.7)

$$S_{ij} = \int d\tau \Phi_i\, \Phi_j$$

(the Φ_i are the determinants of step (ii)).

As we shall see in the next section, the evaluation of the matrix elements (5.3.7) is an almost insuperable problem.

This sketch of the VB method has made no mention of the establishment of spin eigenfunctions before attempting the solution of the linear variation problem and, from one point of view, this "spin theory" is the essence of the VB method: the spin-paired two-electron bond. Our rather cavalier neglect of the finer points of the theory is deliberate on two counts. First, it is our intention to arrive at the central *computational* problems of the VB method as quickly as possible. Second, it is intended to expose clearly that these problems have *not been solved* and indeed are not likely to be solved in the immediate future since they are problems of technique rather than principle. In short, it is not the intention of this analysis to "set VB theory up" for computational implementation. Most of the applications discussed in later chapters are of the MO method. A modified, computationally tractable, form of the VB method is outlined in Chapter 13. The expressions for the VB matrix elements are of general interest in orbital theories (we used one of the simpler results in (4.2.9)) and we now turn to an outline of the derivations of these expressions.

5.4 THE EVALUATION OF VB MATRIX ELEMENTS - THE SLATER/LÖWDIN RULES

The full molecular Hamiltonian contains one-electron and two-electron operators

$$\hat{H} = \sum_{i=1}^{n} \hat{h}(i) + \sum_{i>j=1}^{n} \hat{g}(i,j)$$

Therefore, in the evaluation of matrix elements between determinantal functions

$$\int d\tau\, \Phi_\kappa \hat{H} \Phi_\lambda \qquad (5.4.1)$$

there are two types of term to consider:

$$\int d\tau\, \Phi_\kappa\, \hat{h}(i)\, \Phi_\lambda \tag{5.4.2}$$

and

$$\int d\tau\, \Phi_\kappa\, \hat{g}(i,j)\, \Phi_\lambda \tag{5.4.3}$$

In this section, to avoid a confusing multiple subscript notation, the symbol κ_i will be used for one of the spin-orbitals in Φ_κ and the more usual λ_i for one of the spin-orbitals used in Φ_λ:

$$\begin{aligned} \Phi_\kappa &= \det\{\kappa_1(x_1)\kappa_2(x_2)\ \ldots\ \kappa_n(x_n)\} \\ \Phi_\lambda &= \det\{\lambda_1(x_1)\lambda_2(x_2)\ \ldots\ \lambda_n(x_n)\} \end{aligned} \tag{5.4.4}$$

Each determinant consists of n! orbital-product terms and it is easy to see that the $(n!)^2$ terms defined by the integrals (5.4.2) and (5.4.3) contain many repetitions. Experimenting with 2×2 or 3×3 determinants is enough to suggest the general result that, of the $(n!)^2$ terms, *only n! are distinct*. This means, in practice, that we can choose any *one* of the n! orbital products in Φ_κ and multiply the result by n!. In attempting to evaluate (5.4.2) and (5.4.3) there are two points to consider:

i) How to evaluate the individual orbital product integrals

$$\int d\tau \kappa_1(x_{i_1})\kappa_2(x_{i_2})\ldots\kappa_n(x_{i_n})\hat{H}\lambda_1(x_{j_1})\lambda_2(x_{j_2})\ldots\lambda_n(j_n)$$

arising from the expansion of (5.4.1). $[(i_1,i_2,\ldots i_n)$ and $(j_1,j_2,\ldots j_n)$ are two permutations of $(1,2,\ldots n)]$.

ii) How to collect the large number of such terms together in a final expression for (5.4.1)?

If we expand Φ_κ, Φ_λ and $\hat{H}$ into their explicit forms the smallest elements of this algebraic expression for (5.4.1) are terms like

$$\int d\tau \kappa_1(x_{i_1})\kappa_2(x_{i_2})\ldots\kappa_n(x_{i_n})\hat{h}(i)\lambda_1(x_{j_1})\lambda_2(x_{j_2})\ldots\lambda_n(x_{j_n}) \tag{5.4.5}$$

and

$$\int d\tau \kappa_1(x_{i_1})\kappa_2(x_{i_2})\ldots\kappa_n(x_{i_n})\hat{g}(i,j)\lambda_1(x_{j_1})\lambda_2(x_{j_2})\ldots\lambda_n(x_{j_n}) \tag{5.4.6}$$

Now these multiple integrals (remember $d\tau = dx_1 dx_2 \ldots dx_n$) can be broken down one step further since the operators $\hat{h}(i)$ and $\hat{g}(i,j)$ only involve the coordinates of one or two electrons. Integration over all the electron coordinates except the i'th in (5.4.5) or the i'th and j'th in (5.4.6) can be carried through directly to give a product of overlap integrals over the spin orbitals, each one of the form

$$\int dx \kappa_i(x)\lambda_j(x)$$

The remaining integration over the variable x_i in (5.4.5) gives a one-electron integral of the form

$$\int dx \kappa_k(x_i)\hat{h}(i)\lambda_\ell(x_i)$$

(k and ℓ are determined by the choice of i in (5.4.5)) and the integration over x_i and x_j in (5.4.6) gives an electron repulsion integral, typically,

$$\int dx_i \int dx_j \kappa_k(x_i)\lambda_\ell(x_i)\hat{g}(i,j)\kappa_r(x_j)\lambda_s(x_j)$$

(k, ℓ, r and s determined by i and j).

Thus the "primitive" elements of (5.4.5) and (5.4.6) are products of the familiar AO integrals defined in section 5.2 if we remember that the spatial part of each spin orbital κ_i, λ_i is an atomic orbital. Integrals like (5.4.5) reduce to a product of n-1 overlap integrals and a one-electron integral and integrals like (5.4.6) become a product of n-2 overlap integrals and an electron-repulsion integral. In general, the only terms in this expansion which vanish are ones in which at least one spin integral is

$$\int ds \alpha(s)\beta(s)$$

The orthogonality of the spin functions ensures that such terms are zero independently of the AO's.

The second step, the book-keeping exercise of collecting the ∿n! terms together is best approached by recalling the result (3.3.5) for the normalisation integral of a determinantal function. We can give the overlap determinant involved the symbol $D_{\kappa\kappa}$ where

$$D_{\kappa\lambda} = n! \begin{vmatrix} S_{\kappa_1\lambda_1} & S_{\kappa_1\lambda_2} & \cdots\cdots & S_{\kappa_1\lambda_n} \\ S_{\kappa_2\lambda_1} & S_{\kappa_2\lambda_2} & \cdots\cdots & \\ \cdots & \cdots & \cdots & \cdots \\ S_{\kappa_n\lambda_1} & S_{\kappa_n\lambda_2} & \cdots\cdots & S_{\kappa_n\lambda_n} \end{vmatrix} \qquad (5.4.7)$$

$(S_{\kappa_i\lambda_j} = \int dx\, \kappa_i(x)\lambda_j(x))$

is the general case. This expression is the overlap integral between two *un-normalised determinantal functions;* that is

$$\int d\tau\, \Phi_\kappa \Phi_\lambda = D_{\kappa\kappa}^{-\frac{1}{2}} D_{\lambda\lambda}^{-\frac{1}{2}} D_{\kappa\lambda} \qquad (5.4.8)$$

Consideration of our analysis of the integrals (5.4.4) shows that these integrals are also determinants of spin-orbital integrals. In this case, in every term in the expansion one overlap integral of (5.4.7) is replaced by a one-electron integral. The expansion of the determinant in terms of the one-electron integrals and the n sub-determinants of dimension n-1 gives

$$\int d\tau \Phi_\kappa \sum_{i=1}^{n} \hat{h}(i)\Phi_\lambda = D_{\kappa\kappa}^{-\frac{1}{2}} D_{\lambda\lambda}^{-\frac{1}{2}} \sum_{i,j=1}^{n} D_{\kappa\lambda}(i,j)\int dx\, \kappa_i(x)\hat{h}\lambda_j(x) \qquad (5.4.9)$$

where $D_{\kappa\lambda}(i,j)$ is the determinant obtained from $D_{\kappa\lambda}$ by removing the i'th row (referring to orbital κ_i) and the j'th column (referring to λ_j) and multiplying by $(-1)^{i+j}$ - the ij *cofactor* of $D_{\kappa\lambda}$. The analysis of the two-electron terms is similar but is complicated by the fact that *pairs* of overlap integrals in (5.4.7) are replaced by electron-repulsion integrals. To cut a very long story short, the result is

$$\int d\tau\ \Phi_\kappa \sum_{i>j=1}^{n} \hat{g}(i,j)\Phi_\lambda = D_{\kappa\kappa}^{-\frac{1}{2}} D_{\lambda\lambda}^{-\frac{1}{2}} \sum_{i,j,k,\ell=1}^{n} (\kappa_i\lambda_j,\kappa_k\lambda_\ell)$$

$$\times\ D_{\kappa\lambda}(i,j,k,\ell)\varepsilon_{ik}\varepsilon_{j\ell} \qquad (5.4.10)$$

Here, the cofactor $D_{\kappa\lambda}(i,j,k,\ell)$ is the one obtained by deleting the rows and columns of $D_{\kappa\lambda}$ containing references to orbitals $\kappa_i\kappa_k$ $\lambda_j\lambda_\ell$ respectively. The electron repulsion integral is written in standard notation as $(\kappa_i\lambda_j,\kappa_k\lambda_\ell)$ and

$$\varepsilon_{ik} = 1 \text{ if } i < k$$

or

$$\varepsilon_{ik} = -1 \text{ if } i > k$$

The summation over i j k and ℓ excludes the terms with i=k and j=ℓ.

It is the evaluation of (5.4.10) which prevents the use of the VB method as a routine method in quantum chemistry. There are $\sim n^4$ terms in the summation *for each matrix element* H_{ij}. There is no difficulty in evaluating the cofactors (provided the overlap determinant $D_{\kappa\lambda}$ is non-zero) but there are huge organisational and storage problems associated with the sets of n^4 numbers. In practice the VB method is often formulated in terms of antisymmetric orbital products of the type represented by (5.3.6) since this form has a more transparent chemical interpretation than simple determinantal functions. The expressions for the matrix elements of $\hat{H}$ using these functions are, of course, of similar complexity to (5.4.9) and (5.4.10). The central point from an implementational point of view is the fact that each electron-repulsion integral appears in many matrix elements multiplied by many different coefficients. The advantage of the MO method is that all electron-repulsion effects, for all the MO's, are neatly summarised in G(R).

Equations (5.4.9) and (5.4.10) are completely general in that no assumptions were made about the elements of the overlap determinant $D_{\kappa\lambda}$ and their derivation was first given by P.-O. Löwdin. If the atomic orbitals from which the spin orbitals $\kappa_i\lambda_j$ are formed are *orthogonal* then most of the sub-determinants $D_{\kappa\lambda}(i,j)$ and $D_{\kappa\lambda}(i,j,k,\ell)$ *are zero.* In this case (5.4.9) and (5.4.10) take a particularly simple form due to J.C. Slater. Rather than obtaining Slater's Rules - for matrix elements between determinants of orthogonal spin orbitals - from (5.4.9) and (5.4.10) we will sketch the derivation from first principles.

Returning to (5.4.5), we see that this integral will be zero if any of the component n-1 overlap integrals are zero. Now all the overlap integrals are zero in $D_{\kappa\lambda}$ so the only non-zero terms like (5.4.5) are the ones in which the orbitals on the left are *the same* as the ones on the right except possibly for the orbitals defining the one-electron integral. Thus (5.4.2) is zero unless $\Phi_\kappa = \Phi_\lambda$ or there is only one orbital difference between Φ_κ and Φ_λ. The expressions for these two non-zero integrals are

$$\int d\tau\ \Phi_\lambda \sum_{i=1}^{n} \hat{h}(i)\Phi_\lambda = \sum_{i=1}^{n} \int dx \lambda_i(x)\hat{h}\lambda_i(x)$$

and (5.4.11)

$$\int d\tau\ \Phi_\kappa \sum_{i=1}^{n} \hat{h}(i)\Phi_\lambda = \int dx \kappa_i(x)\hat{h}_j(x)$$

where, in the second case, κ_i and λ_j are the spin orbitals by which Φ_κ and Φ_λ differ. Using the same general approach, it is clear that integrals like (5.4.6) are zero unless all n-2 overlap integrals are non-zero, that is, the orbitals in Φ_κ and Φ_λ have *at most two differences*. The three possible non-zero integrals (5.4.3) are given below. The diagonal element is given by

$$\int d\tau \Phi_\lambda \sum_{i>j=1}^{n} \hat{g}(i,j)\Phi_\lambda = \sum_{i>j=1}^{n} [(\lambda_i\lambda_i,\lambda_j\lambda_j)-(\lambda_i\lambda_j,\lambda_i\lambda_j)] \qquad (5.4.12a)$$

When Φ_κ differs from Φ_λ by the single substitution of κ_i for λ_i then

$$\int d\tau \Phi_\kappa \sum_{i>j=1}^{n} \hat{g}(i,j)\Phi_\lambda = \sum_{j=1}^{n} [(\lambda_i\kappa_i,\lambda_j\lambda_j)-(\lambda_i\lambda_j,\kappa_i\lambda_j)] \qquad (5.4.12b)$$

(excluding the term j=i).
Finally when Φ_κ and Φ_λ differ by two substitutions - κ_i κ_j in place of λ_i λ_j - we have

$$\int d\tau \Phi_\kappa \sum_{i>j=1}^{n} \hat{g}(i,j)\Phi_\lambda = (\lambda_i\kappa_i,\lambda_j\kappa_j)-(\lambda_i\kappa_j,\lambda_j\kappa_i) \qquad (5.4.12c)$$

In constrast to (5.4.9) and (5.4.10) the expressions (5.4.11) and (5.4.12) are very easy to evaluate; most are zero and only the diagonal terms contain a significant number of atomic-orbital integrals. *If the AO's were orthogonal then the VB method would be computationally tractable.* The essentially chemical problem of "choice of structures" would remain but the manipulative use of (5.4.11) and (5.4.12) is quite straight-forward. In fact the AO's are never orthogonal but, as we shall see in Chapter 9, an orthogonal set of orbitals can always be constructed from a non-redundant set of AO's. We shall reconsider the use of the VB method at that time.

From our experience so far in the use of the orbital model the most obvious candidates for a set of orthogonal orbitals are the MO's derived in 5.2. The use of the (orthogonal) *molecular orbitals* within the formalism of VB theory is a theoretically valid procedure and is called

Configuration Interaction (CI) and will be touched on briefly in the closing chapter.

SUGGESTIONS FOR FURTHER READING

Both "The Quantum Theory of Molecules and Solids, I" by J.C. Slater (McGraw-Hill) and "Methods of Molecular Quantum Mechanics" by R. McWeeny and B.T. Sutcliffe (Academic Press) give general discussions of the MO and VB methods. The Roothaan paper referenced in Chapter 4 (Rev.Mod.Phys., 23 69 (1951)) is also relevant here.

A selection of approachs to general VB theory are given below since we shall not take this method very far.

"Quantum Mechanics of Many-Electron Systems" by P.A.M. Dirac Proc.Roy.Soc.(London), A123 714 (1929)

"Quantum Mechanics of Many-Electron Systems in Atoms and Molecules", Chapter 1 of "Tables of Molecular Integrals" by M. Kotani, A. Amemiya, E. Ishiguro and T. Kimura (Maruzen, Tokyo, 1955)

"Spin Free Quantum Chemistry" Part I in Adv.Quant.Chem., 1 (Academic Press 1962) and Part II in J.Phys.Chem., 68 3282 (1964) by F.A. Matsen

"The VB Theory of Molecular Structure, I, II & III" Proc.Roy.Soc. (London), A223 63, 306 (1954) and A224, 288 (1955) discusses the relevance and use of orthogonal orbitals - R. McWeeny.

"Studies in Configuration Interaction I" by I.L. Cooper and R. McWeeny obtains results of a general nature for orthogonal orbitals.

"General Theory of Spin-Coupled Wave Functions for Atoms and Molecules" by J. Gerratt in Adv. in Atomic & Molecular Phys., 7, 47 (1971) is a recent review of VB/CI work.

6 PRACTICAL MOLECULAR WAVE FUNCTIONS

6.1 FURTHER APPROXIMATIONS?

The previous two chapters have defined the chemical and mathematical nature of orbital theories in quantum chemistry and we must now begin to look at the equations involved from a practical, "engineering", point of view. We must ask if the methods we have outlined are feasible computational projects or if further approximations must be made. We may have to "contaminate" our model approximations with approximations of computational convenience. Consideration of the ideas derived in Chapters 4 and 5 from an implementation viewpoint raises questions like: what is the precise nature of the quantities defined by the equations?; how can they be calculated?; what numerical techniques are required?; what is the best way of organising the whole computation? etc. This chapter gives an examination of the effect of such considerations on the choice of basis functions and atomic orbitals.

6.2 MOLECULAR INTEGRAL CONSIDERATIONS

Examination of the definition of the RHF equation (5.2.4) and the Slater/Löwdin rules shows them all to contain integrals involving the AO's (or basis functions) and the one and two electron operators of the molecular Hamiltonian:

$$\int dx_i \mu_i(x_i)\hat{h}(i)\mu_j(x_i) \qquad (6.2.1)$$

and

$$\int dx_i \int dx_j \mu_i(x_i)\mu_j(x_i)\hat{g}(i,j)\mu_k(x_j)\mu_\ell(x_j) \qquad (6.2.2)$$

where the orbitals μ_i may be basis functions χ_i or atomic orbitals ϕ_i. There is a wide variety of rather loose nomenclature for integrals (6.2.1) and (6.2.2). Integrals (6.2.1), which we have called "one electron integrals", are also known as "framework integrals" or "core integrals". Integrals like (6.2.2) are usually called "two electron integrals" or "repulsion integrals". Both types together are often called "molecular integrals". If there are m functions μ_i then, because of the Hermitian symmetry of $\hat{h}$, there are $\frac{1}{2}m(m+1)$ integrals of type (6.2.1) to be evaluated. The "operator" $\hat{g}(i,j)$ is simply a multiplying factor in the integrand and so that the number of distinct integrals (6.2.2) is easily seen to be $\frac{1}{8}(m^4+2m^3+3m^2+2m)$. We shall use the shorthand m$ for this last number.

A specific example is provided by the benzene molecule using m = 36 (5 AO's centred on each carbon atom and 1 on each hydrogen atom). In this case $\frac{1}{2}m(m+1) = 666$ and m$ = 222,111. There are a very large number of molecular integrals, particularly of the type (6.2.2), to be evaluated during an orbital basis valence calculation. We shall therefore need very efficient methods of computing these integrals or we shall have to *choose* our orbitals μ_i so that the integrals can be evaluated rapidly.

In practice, the AO's ϕ_i are expanded in terms of a larger set of basis functions χ_i and the valence calculation is either performed directly in terms of the χ_i or indirectly through the ϕ_i. Thus the number, m, of orbitals in the above example should be the number of basis functions - the *primitive elements* of our orbital model. Atomic calculations have shown that, for elements in the first row of the periodic table, the atomic orbitals can be adequately expressed in terms of about 18 basis functions of STO type (6 s type and 12 p type). This is an average of about three χ's per ϕ. Our benzene example now becomes rather more daunting, 4656 one electron integrals and an overpowering $96\$ = 10841496$ ($\sim 10^7$) electron repulsion integrals. These numbers show that two types of problem are presented by the molecular integrals.

i) How are 10^7 integrals to be computed in a reasonable time? - at first sight it seems scarcely possible to compute 10^7 of *anything* quickly.

ii) Where and how can the computed integrals be stored for future use in the valence calculation?

These two problems are solved by different techniques; one a further approximation of our AO model and the other a computational device.

6.3 APPROXIMATE ATOMIC ORBITALS

In order to preserve the valence analysis of the molecular wave function it is necessary to work with atomic orbitals, but perhaps some practical savings can be made by using "approximate" AO's which are not full solutions of the atomic RHF equations. The approximate AO's can be approximate in two ways; either the expansion length of each AO ϕ_i in terms of the basis functions χ_i can be cut or basis functions can be used which give molecular integrals which are particularly easy to evaluate.

By carefully choosing the orbital exponents ζ in the STO basis functions, it is possible to express each AO approximately as just one term - in this case the distinction between basis functions and atomic orbitals disappears. The radial part of an exact AO and the best one term STO approximation to it is given in Figure 6.1. It can be seen that the main features of the electron distribution are quite well reproduced by the approximate AO. The STO's are particularly suitable for short expansions of AO's because of their similarity to hydrogenic orbitals. Unfortunately, for technical reasons discussed in Chapter 8, the STO's give rise to molecular integrals of the two types (6.2.1) and (6.2.2) which cannot be evaluated by

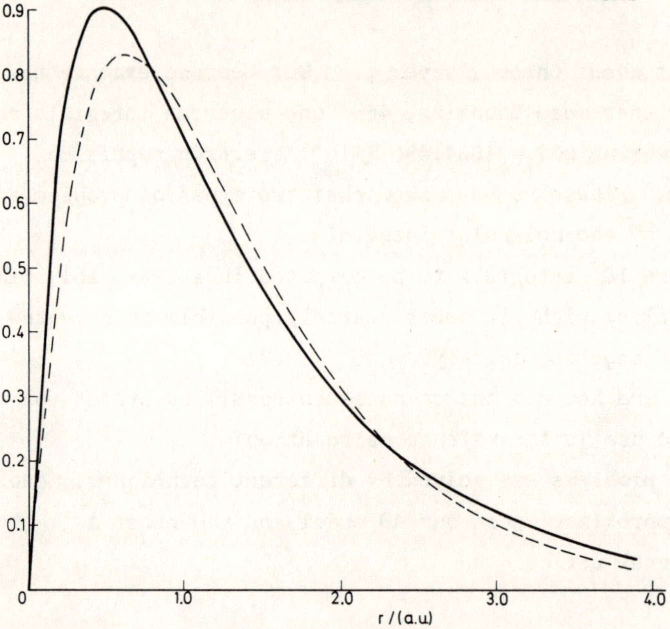

Figure 6.1 The 2p orbital of carbon: full line RHF AO, dotted line single STO

standard analytical techniques and "brute force" methods of direct numerical integration have to be used. Numerical quadrature in many dimensions is particularly time consuming and for all but the largest and most powerful computing facilities (or the smallest molecules!) the routine evaluation of molecular integrals using STO's is out of reach.

We must fall back on the use of atomic basis functions which *we know in advance* give rise to molecular integrals which are analytically tractable. *G*aussian *T*ype *F*unctions (GTF's), which have the general form

$$r^{\nu} \exp(-\alpha r^2) \times (\text{spherical harmonic}) \qquad (6.3.1)$$

do have the property that, when used as basis functions χ_i, they define molecular integrals (6.2.1) and (6.2.2) which are easily evaluated - the most difficult ones reduce to a known function related to the "error function" of probability theory (the derivations are outlined in Chapter 8). The functional form of the GTF's to some extent parallels that of the STO's - the exponential "decay" at large r - and these functions hold out great promise for molecular calculations. The remaining question is "do the GTF's provide a *physically reasonable* basis for the expansion of the AO's?".

Functions very similar to the GTF's are known in quantum mechanics as the eigenfunctions of the linear harmonic oscillator Schrödinger equation

$$(-\tfrac{1}{2}\nabla^2 - \tfrac{1}{2}kr^2)\mu(r) = \varepsilon\,\mu(r) \qquad (6.3.2)$$

(k is the Hooke's law constant).
There are two conclusions which follow from this, first the discouraging one: the potential $-\tfrac{1}{2}kr^2$ is clearly very different from the atomic potential $-Z/r$ and so there is no obvious *physical* reason for the GTF's to form good AO's. On the other hand the fact that GTF's are related to the solutions of the Schrödinger equation (6.3.2) ensures that they can form a complete set and the arguments of 3.8 all hold - any function can be expanded as a linear combination of the solutions of (6.3.2). Thus it is a question of *experiment* to see if the GTF's form a practical set of basis functions.

When GTF's are used in atomic calculations it is found that good approximate AO's can be obtained but the length of the expansion - the number of χ's per ϕ - is greater than for the more physically realistic STO's. Thus the basis of GTF's must be larger than an STO basis for AO's of the same quality - a factor of 3 is typical. Figure 6.2 illustrates a short expansion of the 2s AO of beryllium compared with the accurate AO. The main source of error in the GTF expansion of the atomic orbitals is in the region around the nucleus and hopefully this is a region which is not involved in molecule formation. No very gross errors in molecular electronic density changes will be introduced by using an expansion of the AO's in terms of GTF's.

Experience has shown that the computational simplicity of the GTF molecular integral expressions is the over-riding factor and the use of GTF's now dominates the field of molecular

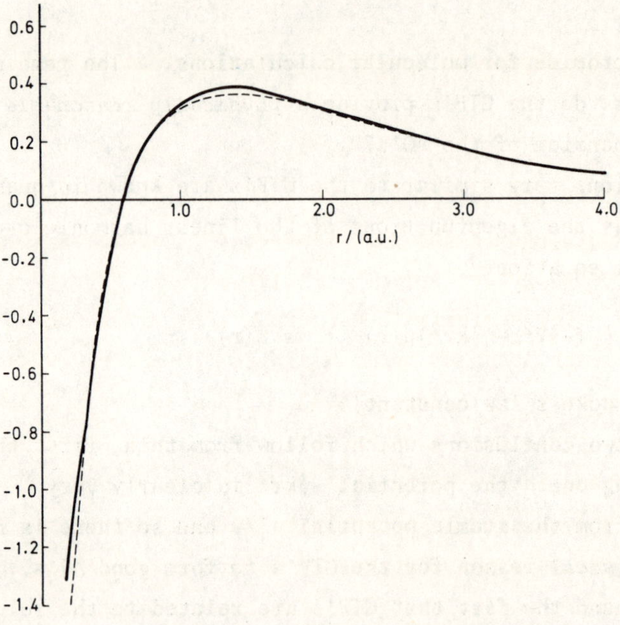

Figure 6.2 The 2s orbital of Beryllium: full line RHF AO, dotted line 2 GTF expansion

valence calculations.

Although the STO's were used as example basis functions in 4.2 the derivation of the RHF equations is quite general and a GTF set of basis functions can be used equally satisfactorily. A GTF expansion of the AO's of a particular atom is obtained by solving the RHF equations while simultaneously varying the orbital exponents - α of (6.3.1) - to obtain a minimum energy for the atom. The optimisation of the so-called non-linear parameters α is a time consuming task; the orbital exponents are varied and for each set of α's the RHF equations solved. The final optimum α's and expansion coefficients are those giving the minimum atomic energy.

An alternative, and more widely used, procedure is to fit a linear combination of GTF's to a known AO (or approximate AO) by, for example, a least squares procedure. An approximate GTF expansion, $\tilde{\phi}$, of an AO ϕ can be written

$$\tilde{\phi} = \sum_{j=1}^{k} C_{ji}\chi_j \qquad (6.3.3)$$

where k is the expansion length. The minimisation of

$$\int dr|\phi - \tilde{\phi}|^2 \tag{6.3.4}$$

with respect to the coefficients C_{ji} of (6.3.3) and the orbital exponents contained in the χ_j optimises the expansion (6.3.3) in the usual least squares sense. A very common "hybrid" of these two ways of using approximate GTF expansions of atomic orbitals is to write each AO as an optimum single STO and to expand this STO in terms of GTF's using (6.3.3). This rather roundabout approach to approximate AO's has its justification in computational convenience and in the historical development of quantum chemistry. Single STO's were used as approximate AO's in setting up molecular wave functions in a qualitative way before accurate computations were technically possible. This familiar, single STO, form of atomic orbitals has "stuck" as a good approximation and, in fact, a beginner may well get the impression that STO's *are* AO's.

The GTF expansion of STO's also has a very valuable convenience property. If the orbital ϕ in (6.3.3) is a STO then, for a given expansion length and orbital type, the optimum coefficients in the expansion are *independent of the orbital exponent of the STO* (ζ). The expansion coefficients only depend on the functional form of the STO - the power of r and the particular spherical harmonic: s, p, d, etc. The surprising fact is that the orbital exponents, α, of the GTF functions χ_j in (6.3.3) are also essentially independent of the STO orbital exponent, the dependence is through a simple scale factor. Thus, the GTF expansion of STO's can be done "once and for all" for a given expansion length k. This expansion method has been carried through for the full range of STO functions using a wide variety of expansion lengths and references to this work are given at the end of this chapter. To show that the orbital exponents of the GTF's in (6.3.3) are independent of STO exponent we can expand (6.3.4) and, if both ϕ and $\tilde{\phi}$ are normalised we have

$$\int dr(\phi - \tilde{\phi})^2 \;=\; 2 - 2\int dr\, \phi\, \tilde{\phi}$$

Using the expansion (6.3.3) for $\tilde{\phi}$,

$$\int dr\, \phi\, \tilde{\phi} \;=\; \sum_{j=1}^{k} C_{ji} \int dr\, \phi\, \chi_j$$

For the simplest "s" type GTF's a typical term in this expansion, for STO exponent unity, is

$$N_G\, N_S \int dr\, r^2 \exp(-\alpha r^2 - r)$$

(where N_G and N_S are the normalising factors of χ_j and ϕ respectively). Transforming variables from r to ζr we have

$$N_G\, N_S \int \zeta^3\, dr\, r^2 \exp(-\alpha\zeta^2\, r^2 - \zeta r)$$

The factors N_G and N_S must both be multiplied by a factor $\zeta^{3/2}$ in order to normalise the GTF $\exp(-\alpha\zeta^2\, r^2)$ and the STO $\exp(-\zeta r)$. Thus the exponents of the GTF's which fit a STO with exponent unity must simply be multiplied by the square of the actual STO exponent to fit that STO.

6.4 CONTRACTION TECHNIQUES

When each AO of a molecular system is expanded in terms of k GTF's, as in (6.3.3), then there are k^4 contributions to an AO electron repulsion integral

$$(\phi_i\phi_j,\phi_k\phi_\ell) \;=\; \sum_{r,s,t,u=1}^{k} C_{ri}C_{sj}C_{tk}C_{u\ell}(\chi_r\chi_s,\chi_t\chi_u) \qquad (6.4.1)$$

to be computed in general. Similarly there are k^2 contributions to the one electron integrals. Equalities among ϕ_i, ϕ_j, ϕ_k, ϕ_ℓ may cut this number in special cases but the factor of k^4 is representative. As we noted in 6.2, if all these GTF integrals are computed and stored, huge amounts of storage space are required. For m AO's each expanded in terms of k

GTF's $(m\ k)\$$ numbers must be stored. If, however, we regard the m AO's as the *essential* degrees of freedom in the calculation - not the mk basis functions - there is no point in storing the GTF integrals seperately since only those combinations forming the AO integrals are ever required. Thus we arrange to compute all the contributions to (6.4.1) consecutively, the summation in (6.4.1) is performed and the whole AO integral is stored as one number. Storage space is then cut to the requirement for $m\$$ electron repulsion integrals. Similar, but less spectacular, savings can be made in the computation of the one electron integrals.

This method of using fixed linear combinations of GTF's and storing only AO integrals is called *contraction* and although mathematically trivial it is of tremendous computational value. Many calculations on the electronic structure of large molecules would be quite impossible without contraction. The use of the contraction technique excludes the second of the possible MO methods described in 5.2 - the use of the basis functions directly in the molecular RHF equation.

6.5 SUMMARY OF APPROXIMATIONS

In view of the number of approximations we have made in developing a computationally feasible theory of molecular electronic structure it is useful at this point to summarise the hierarchy of approximations involved in the MO model since we have now discussed *all* the necessary approximations.

Formally, we shall be working with approximate MO's incompletely expanded in terms of approximate AO's. More optimistically we are forming MO's in a chemically realistic way using AO's which have all the main features of accurate atomic orbitals. In spite of this rather forbidding approximation "tree" we shall continue to use the RHF terminology. For example, the equation

$$H^F T = S T \epsilon$$

will still be called "the molecular RHF equation" and not "the incomplete expression of the molecular RHF equation as provided by the AO's φ expanded in terms of GTF basis χ ".

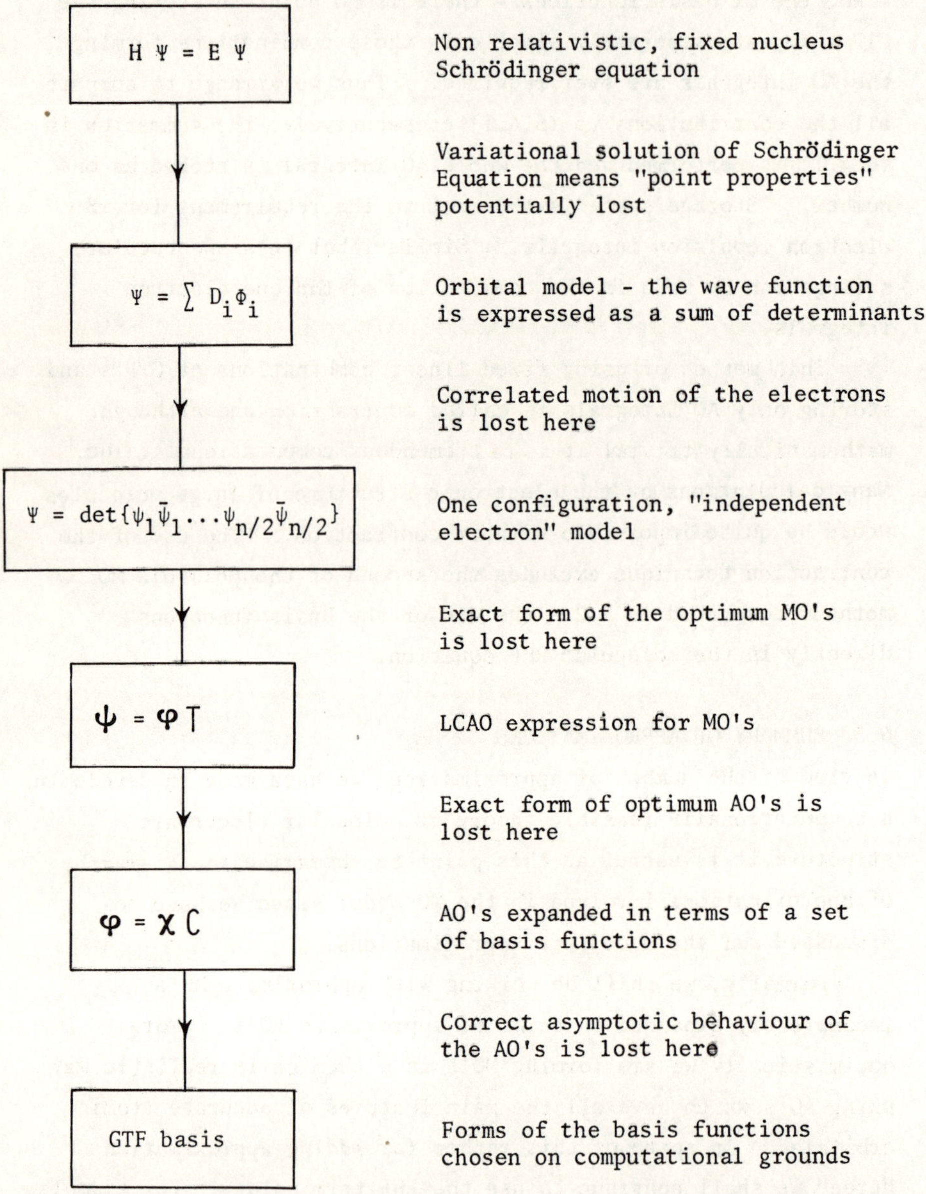

6.6 EXAMPLE - THE BERYLLIUM AND HYDROGEN ATOMS

The development of the central ideas of orbital theories of atomic and molecular electronic structure has, so far, been necessarily rather abstract - summarising and systematising the ideas of qualitative quantum chemistry. It is a very useful aid to understanding these ideas if a specific example is used to illustrate each new stage as it is reached. We are now in a position to begin calculations on a molecular system - defining the AO's and obtaining a GTF expansion of these AO's. At the risk of drawing the chemists' derision we shall work with the (hypothetical) linear BeH_2 molecule. At this stage the separate atoms will be considered. Working with a small molecule has the advantage that the number of molecular integrals, orbital coefficients, etc. is small and these numbers can be conveniently listed for examination and we will not be swamped by numerical detail. It will also become clear that the treatment of BeH_2 shows *all* the features of ab initio valence calculations and, if these are illustrated, the application of the ideas to larger molecules is just a question of computer time.

The ground state 1s AO of the hydrogen atom is known exactly so that two of our atomic orbitals are

$$\begin{aligned} \phi_1 &= N_1 \exp(-|r - R_{H1}|) \\ \phi_2 &= N_1 \exp(-|r - R_{H2}|) \end{aligned} \qquad (6.6.1)$$

where N_1 is the normalising constant 0.564190. The single STO approximate AO's have been computed for the two occupied orbitals of the beryllium atom

$$\phi_3 = N_3 \exp(-3.6848|r - R_{Be}|) \qquad (6.6.2)$$

$$\phi_4 = N_4\, r \exp(-0.9560|r - R_{Be}|) \qquad (6.6.3)$$

The 2p "AO's" of beryllium are not occupied in the ground state of the atom but are sufficiently close in energy to the 2s orbital to be invoked in the qualitative picture of the bonding. The

$2p_z$ orbital - the one along the H-Be-H axis - will be used

$$\phi_5 = N_5 \, z \exp(-0.9560|r - R_{Be}|)$$

In the case of ϕ_5, there is no "atomic reason" for choosing an optimum orbital exponent and the orbital has been given the same exponent as the 2s orbital ϕ_4 by analogy with the hydrogen atom 2s and 2p orbitals. The normalising factors are $N_3 = 3.99067$ $N_4 = 0.29108$ and $N_5 = 0.50416$.

If we now expand each of these AO's in terms of GTF's we obtain a set of approximate AO's suitable for use in the molecular calculation. Table 6.1 gives a selection of expansion coefficients for the orbitals used to describe the electronic structure of the first three rows of the periodic table.

The GTF exponents are listed for a STO orbital exponent of unity and must be multiplied by the square of the actual STO exponent to obtain the best fit for that particular orbital. For our purposes a restricted expansion length of k=2 is adequate - three or four terms is more usual. Using the relevant information from Table 6.1 we obtain the GTF expansion of each of our AO's:

$$\phi_1 = 0.43013\ \chi_1(0.852,0,0,0,r_{H1}) + 0.67891\ \chi_2(0.152,0,0,0,r_{H1})$$

$$\phi_2 = 0.43013\ \chi_3(0.852,0,0,0,r_{H2}) + 0.67891\ \chi_4(0.152,0,0,0,r_{H2})$$

$$\phi_3 = 0.43013\ \chi_5(11.57,0,0,0,r_{Be}) + 0.67891\ \chi_6(2.059,0,0,0,r_{Be})$$

$$\phi_4 = 0.74709\ \chi_7(0.118,0,0,0,r_{Be}) + 0.28560\ \chi_8(0.449,0,0,0,r_{Be})$$

$$\phi_5 = 0.45226\ \chi_9(0.395,1,0,0,r_{Be}) + 0.67131\ \chi_{10}(0.0977,1,0,0,r_{Be})$$

where the extended notation $\chi_i(\alpha,\ell,m,n\ r_A)$ means a normalised GTF

$$N\, x_A^{\ell}\, y_A^{m}\, z_A^{n} \exp[-\alpha|r - R_A|^2]$$

Here,

$$N = \left\{ \frac{2^{2(\ell+m+n)+3/2}\,\alpha^{\ell+m+n+3/2}}{(2\ell-1)!!(2m-1)!!(2n-1)!!} \right\}^{\frac{1}{2}}$$

$$x_A = x - X_A \text{ etc.}$$

$$\text{and } (2\ell-1)!! = 1,3,5,\ldots(2\ell-1)$$

Thus, writing the relation between the basis functions χ_i and the atomic orbitals ϕ_i in the form (4.2.7)

$$\boldsymbol{\varphi} = \boldsymbol{\chi}\,\mathsf{C}$$

then

$$\mathsf{C} = \begin{pmatrix} 0.43013 & 0 & 0 & 0 & 0 \\ 0.67891 & 0 & 0 & 0 & 0 \\ 0 & 0.43013 & 0 & 0 & 0 \\ 0 & 0.67891 & 0 & 0 & 0 \\ 0 & 0 & 0.43013 & 0 & 0 \\ 0 & 0 & 0.67891 & 0 & 0 \\ 0 & 0 & 0 & 0.74709 & 0 \\ 0 & 0 & 0 & 0.28560 & 0 \\ 0 & 0 & 0 & 0 & 0.45226 \\ 0 & 0 & 0 & 0 & 0.67131 \end{pmatrix}$$

Table 6.1 GTF expansions of approximate AO's

GTF expansion for STO 1s, 2s, 3s and 4s in terms of "1s" GTF's

	1s		2s		3s		4s	
k	α	c	α	c	α	c	α	c
2	0.851819	0.430128	0.129228	0.747087	0.669410	-0.152965	0.244179	-0.304666
	0.151623	0.678914	0.0490858	0.285598	0.0583714	1.05137	0.0405110	1.146877
3	2.22766	0.154321	2.58158	-0.0599447	0.564149	-0.178258	0.226794	-0.334905
	0.40577	0.535328	0.156762	0.596039	0.0692442	0.861276	0.0444818	1.05674
	0.1098175	0.444635	0.0601833	0.458179	0.0326953	0.226184	0.0219529	0.125666
4	5.21684	0.0567524	11.6153	-0.0119841	1.51327	-0.0329550	0.324221	-0.112068
	0.954618	0.260141	2.00024	-0.0547205	0.426250	-0.172452	0.166322	-0.284543
	0.265203	0.532846	0.160728	0.580559	0.0764332	0.751851	0.0508110	0.890987
	0.0880186	0.291625	0.0612574	0.477008	0.0376055	0.358963	0.0282907	0.351781

Table 6.1 (continued)

GTF expansion for STO 2p and 3p in terms of "2p" GTF's and STO 3d in terms of "3d" GTF's.

	2p		3p		3d	
k	α	c	α	c	α	c
2	0.432391	0.452263	0.145862	0.534965	0.277743	0.466614
	0.106944	0.671312	0.0566421	0.529961	0.0833651	0.644707
	0.919238	0.162395	2.69288	-0.0106195	0.522911	0.168660
3	0.235919	0.566171	0.148936	0.521856	0.163960	0.584798
	0.0800981	0.422307	0.0573959	0.545002	0.0638663	0.405678
	1.79826	0.0571317	1.85318	-0.0143425	0.918585	0.0579906
4	0.466262	0.285746	0.191508	0.275518	0.292046	0.304558
	0.164372	0.551787	0.0865549	0.584675	0.118757	0.560136
	0.0654393	0.263231	0.0418425	0.214499	0.0528676	0.243242

SUGGESTIONS FOR FURTHER READING

"Atomic Screening Constants from SCF Functions I & II" by E. Clementi & D.L. Raimondi in J.Chem.Phys., 38, 2868 (1963) and 47, 1300 (1967) (with W.P. Reinhherdt) contains optimum single STO AO's for atoms from He through Rn.

"Gaussian Expansion of Slater Type Orbitals" by R.F. Stewart in J.Chem.Phys., 52, 431 (1970) contains GTF expansions of STO's and outlines the least-squares fitting procedure.

7 THE GENERAL STRATEGY

7.1 "SYSTEMS ANALYSIS"

The computation of approximate molecular wave functions is a complex, expensive and time consuming task and before attempting to program any method we must take steps to ensure that we are using the most flexible and general approach, consistent with efficiency. In designing a program structure we must try to anticipate the likely future demands on the system we are developing; the modifications, additions or deletions which other workers will inevitably want to make. We do not want to have to reprogram *anything* unless it is for reasons of numerical efficiency. If it is required to perform a VB calculation in place of the more usual MO type then it is desirable to be able to use as much as possible of the MO program. If different atomic orbitals are to be used or semi-empirical integrals tried or certain integrals neglected or some "inner shell" approximations made, the programs should be sufficiently modular to enable these different approaches to be fitted in.

The importance of designing a flexible program structure *cannot be over-emphasised:* a week spent thinking about the logic involved is worth a month spent writing programs which will be superceded as experience is gained. All too often a researcher spends months developing a suite of programs for a given scientific problem and when his co-workers or successors (or worse - his supervisor) ask "what would we have to do to make these programs do ...?" the answer is "tear the whole thing apart" or "it would be better to start again".

The transition from the derivation of a formal set of equations to the working out of a strategy for their solution often requires a complete change of orientation. As we develop the procedures for handling the numerical quantities involved in the solution of the "orbital model Schrödinger equation" we shall seem very far from the Schrödinger equation as a partial differential equation and, incidentally, still farther from the chemical basis of the mathematics. A diagram of the steps involved in performing molecular valence calculations shows the large-scale modular structure of the computational problem quite clearly - Fig. 7.1 shows the common ground in MO and VB calculations.

7.2 COMPUTATION OF MOLECULAR INTEGRALS

The computation and storage of the molecular integrals H_{ij} and $(ij,k\ell)$ is a necessary first step in any orbital basis calculation, preceding the solution of the matrix equations of either MO or VB type. For a chosen nuclear geometry the same AO integrals can be used for any calculation, within the MO or VB framework, on the electronic structure of the molecule or the related radicals and ions. These integrals must be computed and stored in such a way that they can be kept when the particular calculation on hand is finished. In practice the computation of the AO integrals is the most time consuming step in the calculation and the next chapter is devoted entirely to this subject: efficient computation and flexible storage of the molecular integrals.

As we noted in Chapter 5 the implementation of the VB method is mainly a problem in accessing stored integrals to build up the matrix $\mathbb{H}$ of (5.3.7). The subsequent diagonalisation of $\mathbb{H}$ to solve the linear variational problem is a numerical problem common to both MO and VB methods.

7.3 THE MATRIX LCAO MO EQUATIONS

In our discussion of the atomic Roothaan-Hartree-Fock equations

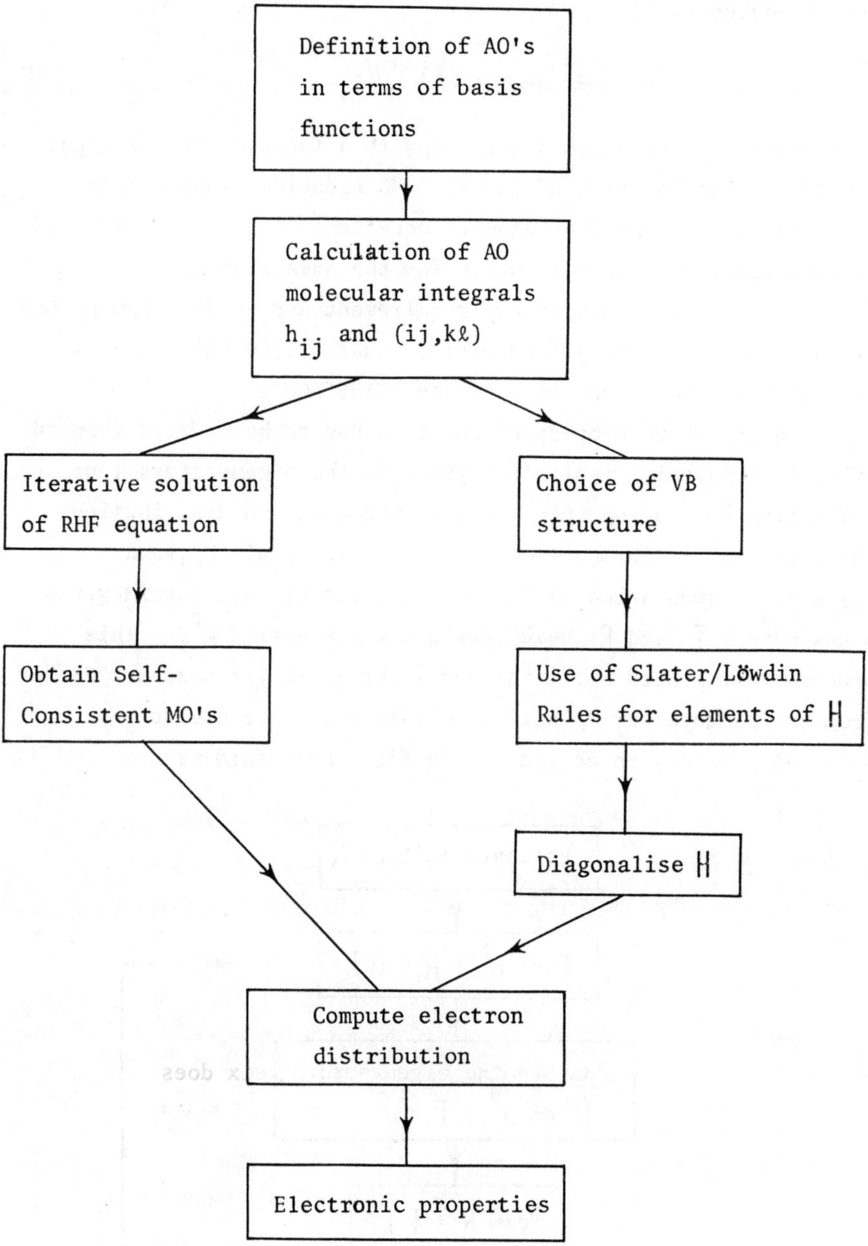

Figure 7.1 The Modular Structure of Molecular Calculations

(Section 4.2) we mentioned that the computation of the eigenvalues and eigenvectors of

$$H^F = H + G(R)$$

is essentially iterative since $G(R)$ is a function of the eigenvectors. The solution of the LCAO MO equation is precisely analogous to that of the "atomic" equations (4.2.21) and is said to be a self-consistent solution for the same reasons. In fact, to stress the self-consistency requirement and to distinguish the eigenvectors of (7.3.1) from earlier qualitative MO's, the molecular orbitals are often called LCAOSCFMO's.

The iterative process of solution has to be started from an initial (physically realistic) guess at the eigenvectors T or the matrix R. Using this guess at the electron distribution the matrix H^F is formed from the stored integrals H_{ij} and $(ij,k\ell)$. Computation of the eigenvectors of this matrix gives a new matrix T (and R) which defines a new matrix H^F. This process is repeated until the new T (or R) matrix does not differ from its predecessor by more than some tolerance decided (on physical grounds) in advance. In flow chart form we have

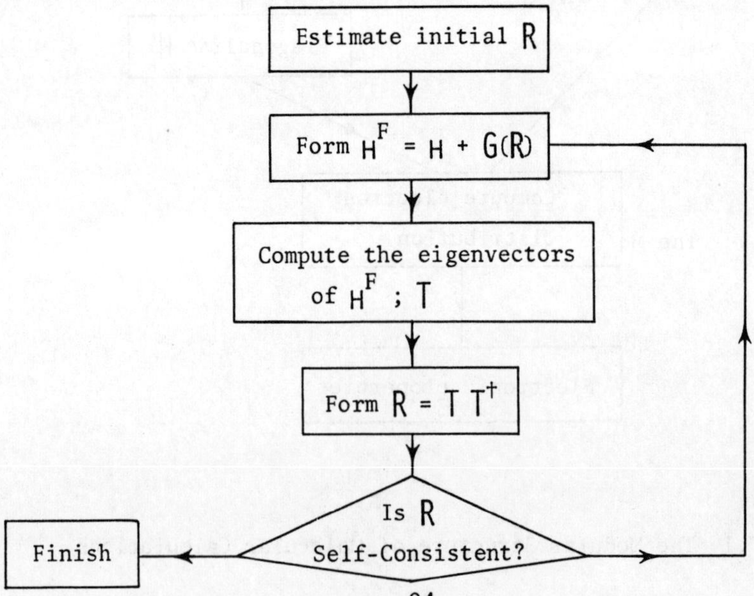

Hopefully, the successive R matrices will converge to a final self-consistent solution rather than oscillating or diverging away from the true solution.

The computational problems in implementing this iterative scheme are:

i) an algorithm for the (efficient) formation of $\mathsf{G(R)}$ from the stored electron repulsion integrals;

ii) a matrix diagonalisation method;

iii) matrix manipulation routines for the addition, multiplication etc. of matrices.

The generation of the matrix $\mathsf{G(R)}$ is an integral part of the method used for storing the electron repulsion integrals and is discussed in Chapter 8. The computation of the eigenvalues and eigenvectors of a symmetric matrix is a problem well researched by applied mathematicians because of its importance in many fields which reduce to linear optimisation. In the next section we give an outline of one "class" of methods because the ideas involved can be applied to optimisation problems other than diagonalisation and are relevant to some of the developments in Chapter 13.

7.4 THE DIAGONALISATION OF SYMMETRIC MATRICES

The matrix LCAOMO problem

$$\mathsf{H}^F \mathsf{T} = \mathsf{T} \epsilon \qquad (7.4.1)$$

where the matrix H^F is m×m and T is m×n/2 (n/2 < m), is an example of the more general matrix eigenvalue problem

$$\mathsf{H}^F \mathsf{U} = \mathsf{U} \epsilon \qquad (7.4.2)$$

where U is m×m. The condition that the columns of U be normalised and orthogonal

$$\mathsf{U}\mathsf{U}^\dagger = \mathsf{1} \qquad (7.4.3)$$

means that U is unitary

$$\mathsf{U}^\dagger = \mathsf{U}^{-1}$$

and so (7.4.2) is equivalent to

$$\mathbb{U}^{\dagger} H^F \mathbb{U} = \boldsymbol{\epsilon} \tag{7.4.4}$$

The problem is thus to compute a matrix $\mathbb{U}$ which transforms H^F into diagonal form via (7.4.4).

Some light can be shed on the general problem by an analysis of the simplest case in which all matrices are 2×2; we require $\mathbb{U}$ such that

$$\begin{pmatrix} U_{11} & U_{21} \\ U_{12} & U_{22} \end{pmatrix} \begin{pmatrix} H^F_{11} & H^F_{12} \\ H^F_{21} & H^F_{22} \end{pmatrix} \begin{pmatrix} U_{11} & U_{12} \\ U_{21} & U_{22} \end{pmatrix} = \begin{pmatrix} \varepsilon_1 & 0 \\ 0 & \varepsilon_2 \end{pmatrix} \tag{7.4.5}$$

Equation (7.4.3), requiring $\mathbb{U}$ to be unitary, reduces the number of independent elements in $\mathbb{U}$ from 4 to 1 since (7.4.3) is, in full,

$$U^2_{11} + U^2_{21} = 1$$

$$U^2_{12} + U^2_{22} = 1$$

$$U_{12}\, U_{11} + U_{22}\, U_{12} = 0$$

$$U_{11}\, U_{12} + U_{21}\, U_{22} = 0$$

These equations can be given a convenient interpretation by considering $\mathbb{U}$ as the matrix defining a rotation in a two dimensional space by an angle θ. Defining

$$c = U_{11} = \cos\theta$$

$$s = U_{12} = \sin\theta \tag{7.4.6}$$

then

$$U_{21} = -s$$

and

$$U_{22} = c$$

satisfy (7.4.3) identically for all θ. We can therefore re-write equation (7.4.5) as

$$\begin{pmatrix} c & s \\ -s & c \end{pmatrix} \begin{pmatrix} H^F_{11} & H^F_{12} \\ H^F_{12} & H^F_{22} \end{pmatrix} \begin{pmatrix} c & -s \\ s & c \end{pmatrix} = \begin{pmatrix} \varepsilon_1 & 0 \\ 0 & \varepsilon_2 \end{pmatrix} \quad (7.4.7)$$

where we have used the fact that $H^F_{12} = H^F_{21}$. Multiplying out the left hand side and collecting the terms defining the (1,2) element of the product matrix gives

$$H^F_{12}(c^2 - s^2) + sc(H^F_{22} - H^F_{11}) = 0 \quad (7.4.8)$$

as a necessary and sufficient condition for the solution of (7.4.7). Defining

$$C = c^2 - s^2 = \cos 2\theta$$

$$S = 2sc = \sin 2\theta$$

and noting $c^2 = \frac{1 + C}{2}$; $s^2 = \frac{1 - C}{2}$

(7.4.8) becomes

$$C\, H^F_{12} + \tfrac{1}{2}S(H^F_{22} - H^F_{11}) = 0$$

$$\text{i.e.} \quad \tan 2\theta = \frac{S}{C} = \frac{2\, H^F_{12}}{H^F_{11} - H^F_{22}} \quad (7.4.9)$$

This is the complete solution of the 2×2 problem since U is defined by θ through (7.4.6).

In general our matrices will be larger than 2×2 and the obvious method to try, having obtained the 2×2 result, is to apply a succession of 2×2 "rotations" to the matrix H^F and eliminate the off-diagonal elements one at a time. A little reflection shows that this is not quite possible because, if our matrix were 4×4, in making the (2,4) element disappear the (1,4) element, earlier transformed to zero, would re-appear. In short the 2×2 rotations *interfere* with each other. The various methods of overcoming this interference define the type of diagonalisation.

i) *The Jacobi method*

This approach uses the optimistic idea that, if all the off-diagonal elements are transformed to zero in turn and sufficiently often, the cumulative effect of the whole set of transformations is to diagonalise the matrix. This proves to be the case - it is simply necessary to keep the process going until all the off-diagonal elements fall below some pre-set tolerance. This method is extremely easy to program, is very stable numerically and is thoroughly well-behaved at degenerate eigenvalues. It is the slowest of the commonly used diagonalisation methods, time consumption being proportional to the fourth power of the size of the matrix.

ii) *The Givens and Householder methods*

It is possible to avoid the interference of the 2×2 rotations if, instead of full diagonalisation, we ask that the matrix be reduced to "tri-diagonal form" - all elements zero except (i,i) and (i,i±1). There are specialised methods for computing the eigenvalues and eigenvectors of tri-diagonal matrices. The Givens method reduces a symmetric matrix to tri-diagonal form by a finite number of 2×2 rotations each of which transforms one off-diagonal element to zero and does not re-introduce other elements previously transformed away. The Householder method uses a more specialised transformation to tri-diagonal form - transforming away a whole column at each rotation. Both of these methods are faster than the Jacobi method - proportional to the cube of the matrix dimension - but are more difficult to program and need special precautions when degenerate eigenvalues occur.

A completely different class of methods - the so-called QR and LU algorithms - are based on the transformation of the matrix to triangular form; references to these methods are

```
      SUBROUTINE JACOB(HF,U,N)
      DIMENSION HF(30,30),U(30,30)
      DATA EPS,ZERO,ONE,TWO,FOUR,BIG/
     * 1.0E-20,0.0,1.0,2.0,4.0,1.0E+20/
C                        SET UP INITIAL UNIT 'U' MATRIX
      DO 1 I=1,N
      DO 2 J=1,N
    2 U(I,J)=ZERO
    1 U(I,I)=ONE
C        HFMAX IS TO BE THE LARGEST HF(I,J)**2
   10 HFMAX=ZERO
C        SWEEP OVER ALL HF(I,J)
      DO 11 I=2,N
      JTOP=I-1
      DO 11 J=1,JTOP
      HFII=HF(I,I)
      HFJJ=HF(J,J)
      HFIJ=HF(I,J)
      HFSQ=HFIJ*HFIJ
C        ENSURE THAT HFMAX IS THE MAXIMUM
      IF(HFSQ.GT.HFMAX)    HFMAX=HFSQ
      IF(HFSQ.LT.EPS)  GO TO 11
C        DO NOT BOTHER WITH ELEMENTS LESS THAN EPS
      DIFFR=HFII-HFJJ
      SIGN=ONE
      IF(DIFFR.GT.ZERO)   GO TO 12
      SIGN=-ONE
      DIFFR=-DIFFR
C                        COMPUTE SIN AND COS VIA TAN
   12 DUM=DIFFR+SQRT(DIFFR*DIFFR+FOUR*HFSQ)
      TAN=TWO*SIGN*HFIJ/DUM
      C=ONE/SQRT(ONE+TAN*TAN)
      S=C*TAN
C        NOW THE EFFECT OF THE CURRENT 2*2 'ROTATION'
      DO 13 K=1,N
      DUM=C*U(K,J)-S*U(K,I )
      U(K,I)=S*U(K,J)+C*U(K,I)
       U(K,J)=DUM
      IF( (I.EQ.K)  .OR. (J.EQ.K) )  GO TO 13
      DUM=C*HF(K,J)-S*HF(K,I)
      HF(K,I)=S*HF(K,J)+C*HF(K,I)
      HF(K,J)=DUM
      HF(I,K)=HF(K,I)
      HF(J,K)=HF(K,J)
   13  HF(I,I)=C*C*HFII+S*S*HFJJ+TWO*S*C*HFIJ
       HF(J,J)=C*C*HFJJ+S*S*HFII-TWO*S*C*HFIJ
      HF(I,J)=ZERO
       HF(J,I)=ZERO
   11 CONTINUE
C        END OF SWEEP - CHECK IF HFMAX IS
C        STILL GREATER THAN EPS
      IF(HFMAX.GT.EPS) GO TO 10
C        HF NOW IN DIAGONAL FORM
C        PUT EIGENVALUES IN ASCENDING ORDER
      DO 20 I=1,N
      HFLIT=BIG
      DO 21 J=I,N
      IF(HFLIT.LT.HF(J,J))  GO TO 21
      HFLIT=HF(J,J)
      JLIT=J
   21 CO NTINUE
      HF(JLIT,JLIT)=HF(I,I)
      HF(I,I)=HFLIT
      DO 20 J=1,N
      ULIT=U(J,JLIT)
      U(J,JLIT)=U(J,I)
   20 U(J,I)=ULIT
      RETURN
      END
```

given in the reading list.

7.5 COMPLICATIONS

We have over-simplified the analysis of the implementation of the LCAOMO method is two respects; no allowance has been made for non-orthogonality of the AO's or of non-convergence of the iterative procedure.

The AO's used in a molecular calculation will never be orthogonal and so the equation to be solved iteratively is, strictly,

$$\mathsf{H}^F \mathsf{T} = \mathsf{S}\,\mathsf{T}\,\boldsymbol{\epsilon}$$

which is not covered by our discussion of matrix diagonalisation methods in the previous section. In fact no new numerical methods are required in the non-orthogonal case, simply three more matrix multiplications during each iteration. The relation between the orthogonal and non-orthogonal basis functions and its application to the solution of the LCAOMO equation using overlapping orbitals is discussed fully in Chapter 9.

In some cases the iterative scheme illustrated in Fig. 7.1 does not converge to a self-consistent T matrix (oscillation rather than divergence is typical). Often these oscillations have a "physical" interpretation - for the current non-self-consistent T matrix there are two possible electronic distributions with very similar energies. The most common solution to this problem is to save a sample of successive T matrices and extrapolate or interpolate from them as iterations are carried out. If two successive matrices are $\mathsf{T}^{(i)}$ and $\mathsf{T}^{(i+1)}$ then the simple average

$$\tfrac{1}{2}(\mathsf{T}^{(i)} + \mathsf{T}^{(i+1)})$$

or a weighted average

$$\mathsf{R}^{(i+1)} + \lambda(\mathsf{R}^{(i)} - \mathsf{R}^{(i+1)})$$

will sometimes induce convergence. If this fails taking the process one step further, using the elements of $\mathsf{T}^{(i+2)}$ in

$$T_{k\ell} = (T_{k\ell}^{(i+2)}T_{k\ell}^{(i)} - T_{k\ell}^{(i+1)2})/(T_{k\ell}^{(i+2)} - 2T_{k\ell}^{(i+1)} + T_{k\ell}^{(i)})$$

gives better convergence behaviour (away from the self-consistent solution where the denominator is not small).

In pathological cases, other techniques based on the direct solution of (5.2.14) and (5.2.15) can be used. Methods of steepest descents and conjugate gradients to obtain the matrix R directly from (5.2.14) have been developed. These methods generally have more well-defined formal convergence properties but are rather unwieldy to implement and are considerably slower in operation than the iterative method for the "normal" well-behaved cases.

7.6 THE VIRTUAL ORBITALS

At the end of the iterative calculation the final self-consistent matrix T contains all the information about the molecular electron distribution within the LCAOMO model. The analysis of the electron density through the matrix R gives the physical and chemical information contained in the single configuration wave function defined by T. Chapter 10 is directed to this analysis. In section 7.4 we found that the diagonalisation of the m×m matrix H^F gives m eigenvectors so the question arises: what interpretation do we put on the m-n/2 eigenvectors contained in U which are not part of T?

The most usual interpretation of these orbitals is based on considerations of a typical one-electron Schrödinger equation such as the hydrogen atom equation

$$\hat{h}\,\phi_i = \varepsilon\,\phi_i \qquad (7.6.1)$$

Here the ϕ_i (and the corresponding ε_i) are associated with the ground and excited states of the atom. By supplying electromagnetic radiation of a suitable frequency the atom can be excited into any of the states whose electron distribution is determined by ϕ_i. The ϕ_i other than the 1s function are then, the excited orbitals of the hydrogen atom. The formal similarity between (7.6.1) and

$$H^F C^{(i)} = \varepsilon_i C^{(i)} \qquad (i > n/2) \tag{7.6.2}$$

has led to a similar interpretation of the "unoccupied" columns of U - the so-called virtual orbitals. That is, the m orbitals in U, when placed in order of increasing ε_i, are all possible levels for the electrons of the molecule to occupy and the ones unoccupied in the ground state determinant are approximations to the excited orbitals. Or, more strictly, a one configuration wave function using one of the virtual orbitals in place of one of the columns of T is an approximation to an excited state of the molecule.

The *formal* similarity of (7.6.2) to (7.6.1) masks the fact that, whereas $\hat{h}$ is independent of the ϕ_i, H^F is a function of T. In the definition of $\mathsf{G(R)}$ the electron repulsion integrals between *each* MO and all the other occupied MO's occur, therefore the virtual orbitals are computed experiencing the electrostatic field of all the occupied orbitals. They are therefore closer to the approximate excited orbitals of the *negative ion* of the molecule in question than to the excited orbitals of the neutral molecule. A harsher view of the virtual orbitals is that they are the residue of all the poor qualities of the AO functions; the variation principle has not acted on them during the iterative solution of the RHF equation. Nevertheless, the virtual orbitals do presumably have the general form of the excited orbitals for a system containing many electrons since one additional electron does not constitute a very severe perturbation to the molecule (c.f. Koopmans' theorem, section 4.3). The virtual orbitals will probably continue to be used as approximate excited orbitals since they have the all-important (computational) advantage of being orthogonal to the occupied molecular orbitals: they are part of the same U matrix.

SUGGESTIONS FOR FURTHER READING

The best references here are the manuals of the program packages listed in Appendix B.

"Mathematical Methods for Digital Computers II" edited by A. Ralston & H.S. Wilf (Wiley 1967) gives a discussion (by B.N. Parlett) of the LU and QR methods for matrix diagonalisation. The same volume contains an article on the Givens-Householder method by J. Ortega.

8 MOLECULAR INTEGRALS - COMPUTATION AND STORAGE

8.1 MOLECULAR INTEGRALS

The computation, storage and retrieval of the AO integrals is the most time consuming part of any orbital basis valence calculation and has therefore been the subject of a large amount of research. The two main areas are the rapid and accurate computation of the integrals and the design of an efficient file structure for storing the computed integrals for subsequent use. The electron repulsion integrals present the most difficulties from both points of view. The computation and storage of the one electron integrals is not critical in view of the relatively small number of them; they can, for example, be comfortably held in the fast store of a computing system. The electron repulsion integrals will, except in the smallest applications, be kept on a file external to the computer's fast store - a disk, drum or magnetic tape. In this chapter we sketch the general problems involved in molecular integral evaluation and describe a storage and retrieval method for the electron repulsion integrals.

8.2 NOTATION

We shall only give the briefest outline of the derivation of the integral formulae in order to show the computational advantages of GTF's over STO basis functions. The complete derivations are lengthy but well known and suitable references are given at the end of this chapter. In any case a full derivation of these formulae would take us too far out of our way and would add nothing to the physical understanding of the various types of integral or to the logic of orbital basis calculations. The operations involved in the integral expressions are

the overlap "operator", 1;

the kinetic energy operator, $\frac{1}{2}\nabla^2$;

the electron-nucleus attraction operator, $1/r_{i\alpha}$;

and the electron repulsion operator, $1/r_{ij}$.

In Fig. 8.1 we give a diagram of an overall ("global") co-ordinate system to which all relative distances can be referred.

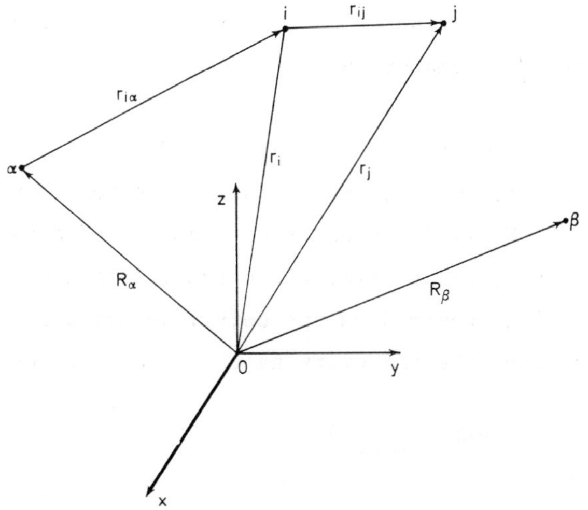

Figure 8.1 Position vectors in the "global" co-ordinate system (i and j are electrons, α and β nuclei)

The commonly occurring relative position vectors are the one between an electron and a nucleus (using upper case symbols for the nuclear position vectors)

$$r_i - R_\alpha = r_{i\alpha}$$

(or simply r_α if only one electron is involved) and an inter-electronic vector

$$r_i - r_j = r_{ij}$$

The moduli of these vectors, representing the relevant distances, are written

$$r_{i\alpha} = |r_{i\alpha}| \text{ (= } r_\alpha \text{ in the case of just one electron)}$$

and

$$r_{ij} = |r_{ij}|.$$

The atomic orbitals and basis functions are centred on the various nuclei and therefore appear, as in Section 6.6, in the global co-ordinate system as functions of $(x_{i\alpha}, y_{i\alpha}, z_{i\alpha})$ and $r_{i\alpha}$. Consideration of the integrand in the simplest type of integral, the overlap, illustrates the main points in the general derivations. The two types of basis function, STO and GTF, are dealt with separately.

8.3 MOLECULAR INTEGRALS USING AN STO BASIS

Working with a particular example shows the general principles quite clearly. We will use our BeH_2 example and consider integrals involving the 1s orbital on one of the hydrogen atoms and the $2p_z$ orbital centred on the beryllium nucleus. Giving these two functions the temporary names χ_{H1} and χ_{Be} we have

$$\chi_{H1} = \exp(-\zeta_{H1} r_{H1})$$

and

$$\chi_{Be} = z_{Be} \exp(-\zeta_{Be} r_{Be})$$

where the normalisation constants have been omitted since they play no rôle in the integration. The overlap integral between the two functions is

$$\int dr\ \chi_{H1}(r)\chi_{Be}(r)$$

$$= \int dx \int dy \int dz (z - Z_{Be}) \exp(-\zeta_{H1} r_{H1} - \zeta_{Be} r_{Be}) \qquad (8.3.1)$$

where

$$r_{H1} = |r - R_{H1}|$$

$$= [(x - X_{H1})^2 + (y - Y_{H1})^2 + (z - Z_{H1})^2]^{\frac{1}{2}}$$

and

$$r_{Be} = [(x - X_{Be})^2 + (y - Y_{Be})^2 + (z - Z_{Be})^2]^{\frac{1}{2}}$$

The three-dimensional integrand in (8.3.1) does not split into separate functions of x, y and z and so the multiple integral does not reduce to a product of three one-dimensional integrals. The square root in the definition of r_{H1} and r_{Be} is the stumbling block in Cartesian co-ordinates. However, a co-ordinate system has been found in which the multiple integrand does reduce to one-dimensional integrations. This is the prolate spheroidal co-ordinate system (ξ,η,ϕ) whose relation to Cartesian and spherical polar systems on each nucleus is given by

$$\xi = \frac{1}{R}(r_\alpha + r_\beta) \ ; \quad \eta = \frac{1}{R}(r_\alpha - r_\beta) \ ; \ \phi = \phi_\alpha = \phi_\beta$$

Hence:

$$r_\alpha = \frac{R}{2}(\xi + \eta) \ ; \qquad r_\beta = \frac{R}{2}(\xi - \eta) \ ;$$

$$\cos\theta_\alpha = \frac{1 + \xi\eta}{\xi+\eta} \ ; \qquad \cos\theta_\beta = \frac{-1 + \xi\eta}{\xi-\eta} \ ; \tag{8.3.2}$$

$$\sin\theta_\alpha = \frac{[(\xi^2-1)(1-\eta^2)]}{\xi+\eta} \ ; \qquad \sin\theta_\beta = \frac{[(\xi^2-1)(1-\eta^2)]}{(\xi-\eta)}$$

$$x_\alpha = x_\beta = \frac{R}{2}[(\xi^2-1)(1-\eta^2)]^{\frac{1}{2}}\cos\phi; \ y_\alpha = y_\beta = \frac{R}{2}[(\xi^2-1)(1-\eta^2)]^{\frac{1}{2}}\sin\phi$$

$$z_\alpha = z_\beta + R = \frac{R}{2}(1+\xi\eta) \ ; \qquad 1 \leqslant \xi \leqslant \infty, \quad -1 \leqslant \eta \leqslant 1.$$

The volume element dV is given by

$$dV = \left(\frac{R}{2}\right)^3 (\xi^2 - \eta^2)d\xi d\eta d\phi$$

and

$$R = |R_\alpha - R_\beta|$$

Surfaces of constant ξ are confocal prolate spheroids and surfaces of constant η are confocal hyperboloids. The variables in relations (8.3.2) are illustrated in Fig. 8.2.

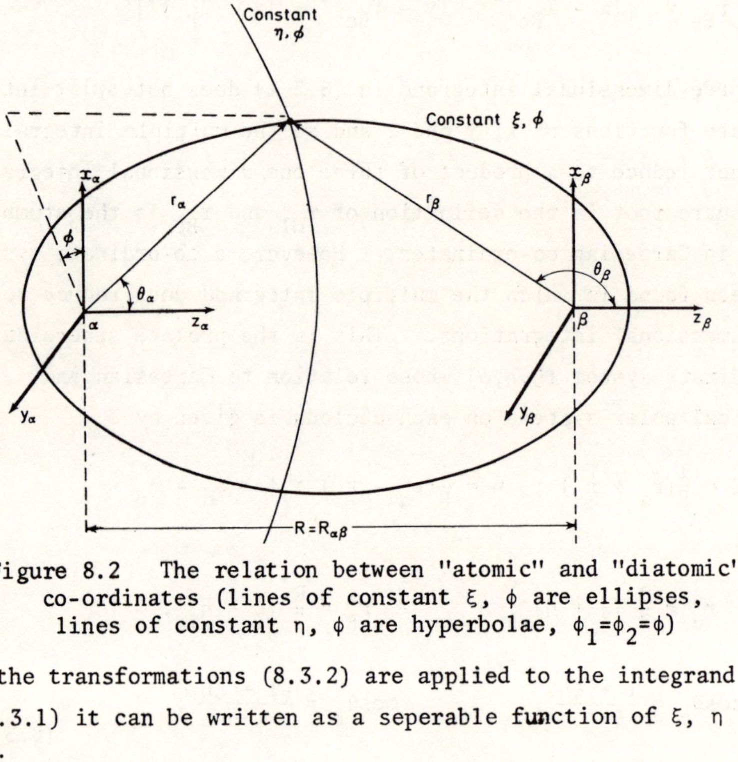

Figure 8.2 The relation between "atomic" and "diatomic" co-ordinates (lines of constant ξ, ϕ are ellipses, lines of constant η, ϕ are hyperbolae, $\phi_1=\phi_2=\phi$)

When the transformations (8.3.2) are applied to the integrand in (8.3.1) it can be written as a seperable function of ξ, η and ϕ:

$$\left(\frac{R}{2}\right)^4 \int_1^\infty d\xi \int_{-1}^{1} d\eta(\xi\eta-1)(\xi^2-\eta^2)\exp[-p(\xi+t\eta)] \int_0^{2\pi} d\phi$$

where $p = \frac{R}{2}(\zeta_{H1} + \zeta_{Be})$

and $t = (\zeta_{H1} - \zeta_{Be})/(\zeta_{H1} + \zeta_{Be})$

Each of the separated integrals can be evaluated analytically since they are known standard forms. Although our argument has been developed for two particular STO functions, reference to the general form of the STO's shows that the reduction can be carried through in all cases and all STO overlap integrals can be evaluated analytically. The two types of standard form to which all these overlap integrals reduce are the so-called A and B functions;

$$A_n(x) = \int_1^{\infty} dt\ t^n \exp(-tx)$$

and

$$B_n(x) = \int_{-1}^{1} dt\ t^n \exp(-tx) = -A_n(x) - (-1)^n A_n(-x)$$

The expressions used for the numerical computation of these standard integrals are

$$A_n(x) = \frac{n!\exp(-x)}{x^{n+1}} \sum_{k=0}^{n} \frac{x^k}{k!} \qquad (\rho>0)$$

$$= \frac{1}{x}\,[\exp(-x) + n\,A_{n-1}(x)]$$

and

$$B_n(x) = 2\sum_{i=0}^{\infty} \frac{x^{2i}}{(2i)!\,(n+2i+1)} \;; \qquad n \text{ even}$$

$$= -2\sum_{i=1}^{\infty} \frac{x^{2i-1}}{(2i-1)!\,(n+2i)} \;; \qquad n \text{ odd}$$

The general analysis of the STO kinetic energy integrals is rather more involved because of the differentiations occurring in the integrand. It can be readily seen, however, that differentiating a STO produces another STO-like function:

$$\frac{\partial}{\partial x}\exp(-\zeta r) = \frac{\partial}{\partial x}\exp[-\zeta(x^2 + y^2 + z^2)^{\frac{1}{2}}]$$

$$= \frac{-x}{r}\exp(-\zeta r).$$

Performing all the differentiations implied by ∇^2 shows that it is possible to reduce the kinetic energy integrals to a weighted sum of STO overlap integrals. Thus, the STO kinetic energy integrals can all be evaluated using the standard A and B functions.

The electron-nucleus attraction integrals present a problem which also occurs in the analysis of the electron repulsion integrals. In our example the attraction energy between the electron density $\chi_{H1}\chi_{Be}$ and the second hydrogen nucleus is given by

$$\int dx \int dy \int dz \; \chi_{H1}\left(-\frac{1}{r_{H2}}\right)\chi_{Be}$$

$$= -\left(\frac{R}{2}\right)^4 \int_1^\infty d\xi \int_{-1}^{1} d\eta \int_0^{2\pi} d\phi\,(\xi\eta-1)(\xi^2-\eta^2)\exp[-p(\xi+t\eta)]\frac{1}{r_{H2}}$$

when the orbital product part of the integrand has been expressed directly in the (ξ,η,ϕ) system. Unfortunately, the expression for r_{H2} does *not* separate into individual functions of ξ η and ϕ. In a molecule of general geometry this is true of the distances from all nuclei other than the two on which the orbitals in the integrand are centred. In fact,

$$r_{H2}^2 = r_{H1}^2 + R_{H2}^2 - 2r_{H1}.R_{H2}$$

$$= 2\left(\frac{R}{2}\right)^2 [Q - S\cos(\phi-\phi_{H2})]$$

where

$$Q = \xi^2 + \xi_{H2}^2 + \eta^2 + \eta_{H2}^2 - 2\xi\xi_{H2}\eta\eta_{H2} - 2$$

$$S = 2[(\xi^2-1)(1-\eta^2)(\xi_{H2}^2-1)(1-\eta_{H2}^2)]^{\frac{1}{2}}$$

(ξ,η,ϕ) are the co-ordinates of the electron

and $(\xi_{H2},\eta_{H2},\phi_{H2})$ those of the nucleus.

So, except for diatomic molecules (and atoms), the "three centre nuclear attraction integral problem" prevents the full analytical evaluation of even the one electron Hamiltonian integrals for STO basis functions.

The electron repulsion integrals are physically and mathematically very similar to the nuclear attraction integrals - they represent the interaction between two electron distributions instead of between an electron distribution and a point charge. There is an additional integration to be carried out over the three spatial co-ordinates of the second electron. Clearly if the general nuclear attraction integral cannot be evaluated analytically, the electron repulsion integrals will be analytically intractable. In this case, the inter-electronic distance is given by

$$r_{12}^2 = r_{1\alpha}^2 + r_{2\alpha}^2 - 2r_{1\alpha}\cdot r_{2\alpha}$$

$$= 2\left(\frac{R}{2}\right)^2 [Q - S\cos(\phi_1 - \phi_2)]$$

where $Q = (\xi_1^2 + \xi_2^2 + \eta_1^2 + \eta_2^2 - 2\xi_1\xi_2\eta_1\eta_2 - 2)$

$$S = 2[(\xi_1^2-1)(1-\eta_1^2)(\xi_2^2-1)(1-\eta_2^2)]^{\frac{1}{2}}$$

(ξ_1,η_1,ϕ_1) and (ξ_2,η_2,ϕ_2) are the co-ordinates of the two electrons.

Certain special cases of the electron repulsion integrals - the "Coulomb integrals" can be evaluated exactly. That is, electron repulsion integrals of the form

$$\int dr_1 \int dr_2 \chi_i^\alpha(r_1)\chi_j^\alpha(r_1) \frac{1}{r_{12}} \chi_k^\beta(r_2)\chi_\ell^\beta(r_2)$$

where χ_i^α, χ_j^α are both centred on nucleus α and χ_k^β, χ_ℓ^β are on nucleus β, can be evaluated analytically. The remaining "three and four centre" integrals require separate treatment. In general, evaluation of an electron repulsion integral involving a "two-centre" charge distribution $\chi_i^\alpha\chi_j^\beta$ $(\alpha\neq\beta)$ involves numerical integration. In certain special cases, notably diatomics and other linear molecules, the high symmetry of the molecule enables certain integrations to be obtained in closed form (the ϕ integrations) leaving a multiple integral of smaller dimension to be evaluated numerically.

The general polyatomic molecule requires numerical integration techniques to be used for the bulk of the STO molecular integrals. Numerical quadrature over 3, 4 or 6 dimensions is a very time consuming process; typically involving the computation of the integrand at least 10^3, 10^4 or 10^6 times respectively for reliable integral values. Bearing in mind the need to compute perhaps hundreds of thousands of these integrals, this is a problem to be tackled only by those with access to the largest and fastest computing facilities. We conclude then, that the *routine* use of STO's for the calculation of molecular wave functions is restricted to linear molecules (and atoms).

As we have seen, it is the fact that the distances r_{H1} and r_{Be} occur *as such* in the exponentials of the integrand which results in the molecular integrals being irreducible in Cartesian co-ordinates. If r_{H1} and r_{Be} occur as r_{H1}^2 and r_{Be}^2 then the exponentials separate in Cartesian co-ordinates and, for example, the overlap integrals become products of three one dimensional integrals over x, y and z. This is one of the advantages of the use of GTF basis functions.

8.4 MOLECULAR INTEGRALS USING A GTF BASIS

The GTF analogues of the 1s and $2p_z$ functions we used in the last section are

$$\chi_{H1} = \exp(-\alpha_{H1} r_{H1}^2)$$

and

$$\chi_{Be} = z_{Be} \exp(-\alpha_{Be} r_{Be}^2)$$

(again dropping the normalisation factors). The overlap integral between these two functions is

$$\int dx \int dy \int dz (z - Z_{Be}) \exp(-\alpha_{H1} r_{H1}^2 - \alpha_{Be} r_{Be}^2)$$

where r_{H1} and r_{Be} are given by the expressions in the last section. The exponential term in the integrand can be simplified by the fact, familiar from statistics, that the product of two normal (Gaussian) distributions is normal. That is, *the product of two 1s GTF's is a third 1s GTF centred on a point on the line joining the centres of the original GTF's.* If

$$\mathrm{P} = \frac{\alpha_{H1}\mathrm{r}_{H1} + \alpha_{Be}\mathrm{r}_{Be}}{\alpha_{H1} + \alpha_{Be}}$$

$$\mathrm{r}_p = \mathrm{r} - \mathrm{P}; \qquad r_p = |\mathrm{r}_p|$$

$$\mathrm{R} = \mathrm{r}_{H1} - \mathrm{r}_{Be}; \qquad R = |\mathrm{R}|$$

then

$$\exp(-\alpha_{H1}r_{H1}^2 - \alpha_{Be}r_{Be}^2) = \exp\left(\frac{-\alpha_{H1}\alpha_{Be}R^2}{\alpha_{H1}+\alpha_{Be}}\right)\exp[-(\alpha_{H1}+\alpha_{Be})r_p^2] \tag{8.4.1}$$

where the first exponential on the right-hand side is a constant and takes no further part in the analysis and so will be dropped. Since the remaining term in the integrand can be expressed in terms of the components of r_p;

$$(z - Z_{Be}) = z_p + (\mathrm{P} - \mathrm{R}_{Be})_x = z_p + C \quad \text{(say)}$$

the whole integrand can be expressed in terms of r_p as independent variable. Performing this transformation gives

$$\int dx_p \int dy_p \int dz_p \; (z_p + C) \exp(-\alpha r_p^2)$$

where $\alpha = \alpha_{H1} + \alpha_{Be}$.

We can now separate the integral in Cartesian coordinates giving

$$\int dx_p \exp(-\alpha x_p^2) \int dy_p \exp(-\alpha y_p^2) \int dz_p (z_p + C)\exp(-\alpha z_p^2)$$

The integration limits for each of the coordinates x, y and z are, of course, ±∞. Thus, for the GTF overlap integral, the basic integrals

$$\int_{-\infty}^{\infty} dt\ t^n \exp(-\alpha t^2)$$

have to be evaluated. Only the integrals with even n are non-zero and

$$\int_{-\infty}^{\infty} dt\ t^{2n} \exp(-\alpha t^2) = \frac{(2n-1)!!}{2^n} \sqrt{\frac{\pi}{\alpha^{2n+1}}}$$

The kinetic energy integrals are reducible to a weighted sum of overlap integrals since the derivative of a GTF is another GTF-like function:

$$\frac{\partial}{\partial x} (\exp -\alpha r^2) = -2\alpha x \exp(-\alpha r^2)$$

and, just as we found in the case of STO kinetic energy integrals, no new analysis is required.

The analysis of the nuclear attraction and electron repulsion integrals is simplified by the property (8.4.1) of GTF's. Further, the inverse distance operators can be expressed in a "Gaussian-like form" (at the expense of one additional integration) by the transformation

$$\frac{1}{r_{i\alpha}} = \frac{1}{\sqrt{\pi}} \int_0^{\infty} ds\ s^{-\frac{1}{2}} \exp(-s r_{i\alpha}^2)$$

and

$$\frac{1}{r_{ij}} = \frac{1}{\sqrt{\pi}} \int_0^{\infty} ds\ s^{-\frac{1}{2}} \exp(-s r_{ij}^2)$$

Remembering that $r_{i\alpha}^2$ and r_{ij}^2 split into separate functions of the Cartesian co-ordinates, these transformations enable the nuclear attraction integrals to be split into three one-dimensional integrals over x, y and z and an additional "outer" integration over the transformation dummy variable s. The electron repulsion integrals similarly become six separate Cartesian integrations and the integral over s. The final integration over s reduces to the standard form

$$\int_0^1 ds\ s^{2\nu} \exp(-ts^2) = F_\nu(t)$$

This function, $F_\nu(t)$, is related to the "error function" of statistics and methods for its evaluation are well known.

This brief outline shows only the main features of the GTF integral derivation and working formulae for each type of integral are given in Appendix A together with references to the full derivation.

As we have seen, the GTF functions have the important property of allowing the various molecular integrals to be separated in Cartesian co-ordinates. It is therefore convenient to use a Cartesian-orientated form for the GTF functions themselves; the expression used in 6.6;

$$N\ x_A^{\ell}\ y_A^{m}\ z_A^{n} \exp(-\alpha r_A^2)$$

is commonly used in place of the spherical harmonic form of (6.3.2). The Cartesian and spherical harmonic forms of the GTF basis functions are very simply related; the s and p types are identical and the higher forms are simple linear combinations:

$$d_{3z^2-r^2} = d_{z^2} - \tfrac{1}{2}(d_{x^2} + d_{y^2})$$

$$d_{x^2-y^2} = \frac{\sqrt{3}}{2}\ (d_{x^2} - d_{y^2}).$$

In order to retain the atomic orbital nomenclature, the GTF's are classified by the sum of the exponents of x_A, y_A and z_A:

$\ell+m+n = 0$ is an "s" type GTF

$\ell+m+n = 1$ is a "p" type GTF

$\ell+m+n = 2$ is a "d" type GTF

etc.

The numerical evaluation of the expressions given in Appendix A is very dependent on the values of ℓ, m and n of the orbitals in the integrand. Integrals involving s-type orbitals can be evaluated extremely rapidly. The p and d orbitals involve integrals which are rather slower to evaluate and higher values of ℓ+m+n ("f", "g", etc. orbitals) give molecular integrals which are prohibitively time-consuming for most applications. Many molecules of chemical interest involve bonding among s, p and d orbitals and the computation of molecular integrals for these molecules using a GTF basis is quite a straightforward task.

From a practical point of view it is often found to be rather wasteful to program the expressions of Appendix A in full generality if the resultant program is to be used, for example, mostly for s type and p type integrals. It is certainly worth-while to write at least one separate program for integrals involving all s-type GTF's and arrange for this special program to be used in the appropriate cases. Many workers write separate programs for molecular integrals involving each combination of s, p, d GTF's and try to optimise each routine for good performance in that particular area.

The equations of Appendix A apply to the GTF basis functions, of course, and the atomic orbital integrals are obtained by summing the GTF contributions to each AO integral - the contraction technique of 6.4.

8.5 PHYSICAL INTERPRETATION AND ORDERS OF MAGNITUDE

The "coulomb terms" in the molecular Hamiltonian give rise to orbital basis molecular integrals which have a direct electrostatic interpretation and which suggest the physical interpretation of the kinetic energy integrals. Recalling the probability density interpretation of the orbital products,

$$P_{ij}(r) = \chi_i(r)\chi_j(r) \tag{8.5.1}$$

is the probability density of an electron confined to the region of overlap between χ_i and χ_j. The integral of (8.5.1) over all space is a measure of how strongly the two orbitals overlap. The electron-nucleus attraction integrals

$$\int dr\ \chi_i(r)\ -\frac{z_\alpha}{r_\alpha}\ \chi(r) \ = \ -\ z_\alpha \int dr \frac{P_{ij}(r)}{r_\alpha} \qquad (8.5.2)$$

are simply the Coulomb energy of a negative charge distribution $P_{ij}(r)$ in the field of a positive point charge z_α. The electron repulsion integrals

$$\int dr_1 \int dr_2 \chi_i(r_1)\chi_j(r_1)\ \frac{1}{r_{12}}\ \chi_k(r_2)\chi_\ell(r_2)$$

$$= \int dr_1 \int dr_2 P_{ij}(r_1)\ \frac{1}{r_{12}}\ P_{k\ell}(r_2) \qquad (8.5.3)$$

represent the mean repulsion energy between two negative charge distributions $P_{ij}(r_1)$ and $P_{k\ell}(r_2)$. Rearrangement of the kinetic energy integrals into this probability density form is not possible because of the essential *operator* nature of ∇^2. However, by analogy with the other molecular integrals, we interpret the kinetic energy integral

$$\int dr\ \chi_i(r)(-\tfrac{1}{2}\nabla^2)\chi_j(r) \qquad (8.5.4)$$

as the mean kinetic energy of a distribution of charge given by $P_{ij}(r)$.

This interpretation of the molecular integrals enables a few words to be said about the relative magnitudes of some of them. Clearly if the centres of the two functions χ_i and χ_j are widely separated, the exponential form of the χ's ensures that the product $\chi_i\chi_j$ will be small everywhere. In this case any integrals in which $\chi_i\chi_j$ appears as a factor in the integrand will be small. The overlap integral - the sum of the distribution - will be small and in a certain sense this overlap integral measures "how much of an electron is contained in $\chi_i\chi_j$". The Coulombic integrals (8.5.2) and (8.5.3) and the kinetic energy integrals are usually small when the overlap integral involving the same orbitals is small. We can therefore expect that many of the electron repulsion integrals - which have two P_{ij} factors in the integrand - will be small since, in a typical molecule, there are many distant pairs of orbitals.

The overlap integral may also be small or zero for reasons of symmetry: the positive and negative regions of overlap may cancel out. In these cases we can say nothing in general about the sizes of the integral in which the product $\chi_i\chi_j$ appears as a factor. An important example is provided by s and p orbitals on the same centre; here the s-p overlap integral is zero by symmetry and yet the electron repulsion integral

$$\int dr_1 \int dr_2\ s(r_1)p(r_1)\frac{1}{r_{12}}\ s(r_2)p(r_2)$$

is often large since it is the repulsion of an electron density "with itself".

The idea that the overlap density (or "differential overlap") governs the relative sizes of the various electron repulsion integrals is the basis of many approximation methods within the MO framework. The size of the overlap density is used as a criterion for the systematic neglect of classes of electron repulsion integral. Unfortunately, there are very many of the very small electron repulsion integrals and it is difficult or impossible to say, a priori, what effect their neglect has on an MO calculation. Frequently the size of the overlap *integral* is used as a criterion rather than the actual overlap density; a better criterion, avoiding cancellation errors, would be

$$\int dr|\chi_i(r)||\chi_j(r)|$$

An important limiting case for the form of the electron repulsion integrals occurs when two functions χ_i, χ_j are centred on one nucleus, α say, and χ_k, χ_ℓ are centred on nucleus β. Then, if α and β are sufficiently far apart, $r_{12} \approx |R_\alpha - R_\beta|$ and

$$\int dr_1 \int dr_2\, \chi_i(r_1)\chi_j(r_1)\,\frac{1}{r_{12}}\,\chi_k(r_2)\chi_\ell(r_2)$$

$$\approx \frac{1}{|R_\alpha - R_\beta|}\int dr_1\, \chi_i(r_1)\chi_j(r_1)\int dr_2 \chi_k(r_2)\chi_\ell(r_2)$$

$$= \frac{S_{ij}\;\; S_{k\ell}}{|R_\alpha - R_\beta|} \qquad \left(= \frac{1}{|R_\alpha - R_\beta|} \text{ if } i=j,\ k=\ell\right)$$

In the same way we can reduce the two centre nuclear attraction integral:

$$\int dr\, \chi_i(r) - \frac{Z_\gamma}{r_\gamma}\,\chi_j(r) \approx - \frac{Z_\gamma\;\; S_{ij}}{|R_\alpha - R_\gamma|}$$

since $r_\gamma \approx |R_\alpha - R_\gamma|$ if nuclei α and γ are sufficiently well separated. These very coarse considerations lead us to expect that the electron repulsion integrals containing one or more "one centre" charge distributions $\chi_i^{\alpha}\chi_j^{\alpha}$ will be large compared to those containing only "two centre" terms $\chi_i^{\alpha}\chi_j^{\beta}$.

8.6 THE COMPUTER STORAGE OF MOLECULAR INTEGRALS - CONVENTIONS

The matrix of one electron integrals can be conveniently stored in a two-dimensional array in the fast store of a computer. If space saving is important the distinct elements only can be stored, reducing storage to $\frac{1}{2}m(m+1)$ locations. The treatment of the four-index electron repulsion integrals is quite different. The obvious method of using a four-dimensional array in the fast store is out of the question for two reasons:

i) Such an array would have m^4 elements, for m AO's, and this would permit the treatment of only the smallest molecules
ii) There are important equalities among the integrals $(ij,k\ell)$ which permit considerable space saving - m $\not{8}$ in place of m^4 locations.

The second point, due to the permutations of the indices i, j, k and ℓ, is independent of molecular symmetry and so will occur in all cases. Thus we need only compute and store *one* of $(ij,k\ell)$, $(ji,k\ell)$, $(ij,\ell k)$, $(k\ell,ij)$, $(\ell k,ij)$, $(k\ell,ji)$, $(\ell k,ji)$, $(ji,\ell k)$. In using the stored integrals we arrange to ensure that these equalities are used, There are 1, 2, 4 or 8 integrals with the same value as $(ij,k\ell)$, depending on equalities among i, j, k and ℓ.

Having decided to work with only one of the equivalent AO integrals, we must decide *which one* to store, that is, an *ordering convention* among i, j, k and ℓ. There are, of course, just two possibilities and both are used:

i) $i \geqslant j$; $k \geqslant \ell$; $[ij] \geqslant [k\ell]$

ii) $i \leqslant j$; $k \leqslant \ell$; $[ji] \leqslant [\ell k]$

where $[ij] = \frac{1}{2}i(i-1)+j$. There is no pressing reason to choose either: we shall use the first. Thus of the four equivalent electron repulsion integrals

(2 2, 2 1) ; (1 2, 2 2) ; (2 1, 2 2) and (2 2, 1 2)

the first one, which satisfies ordering convention (i), will be used for the whole set of integrals equal to

$$\int dr_1 \int dr_2 \phi_i(r_1)\phi_j(r_1) \frac{1}{r_{12}} \phi_k(r_2)\phi_\ell(r_2)$$

When these integrals are used in, for example, a MO calculation any need for the integral (2 1, 2 2) or (2 2, 1 2) will be met by using (2 2, 2 1).

The decision to store the electron repulsion integrals on an external file presupposes that we shall compute the integrals in blocks of less than m\$ and write these blocks to the file until the whole m\$ have been computed and stored. This enables the integrals to be processed by subsequent programs in manageable amounts. Obviously the simplest thing to do is to compute the distinct integrals one at a time and write each one to the file as it is computed. This is not practical for technical reasons: it is extremely wasteful to ask the computer operating system to hold a program and set up the appropriate channels to a disk file for the transfer of one number $m^4/8$ times. Access to external files is very slow compared to processing time and so the integrals must be stored in blocks which are as large as practicable. The steps in the process are, in outline:

i) assign a block of storage in the fast store - a "buffer" - for temporary storage of the integrals;

ii) set up loops ranging over i, j, k and ℓ satisfying the ordering convention $i \geqslant j$; $k \geqslant \ell$; $[ij] \geqslant [k\ell]$;

iii) using the current labels i, j, k, ℓ and the molecular geometry, orbital specifications, etc., compute the current integral $(ij,k\ell)$;

iv) add the computed $(ij,k\ell)$ to the buffer; if this fills the buffer go to (v), if not go to (vi);

v) write the contents of the buffer to disk;

vi) increment i, j, k, ℓ consistent with the ordering convention of (ii);

vii) if the loops on i, j, k, ℓ are exhausted - i.e. $\ell = m$, go to (viii), if not go to (iii);

viii) close the integral storage file and finish.

When the computed integrals have been stored we need to access and identify them for use in the orbital basis calculation. There are two general ways of arranging for the unique identification of an integral in such a file:

i) The values of the labels i, j, k, ℓ are *stored in the file* together with the value of the integral $(ij,k\ell)$

ii) The integrals $(ij,k\ell)$ are stored *in a particular order* in the file so that the values of the labels i, j, k, ℓ are determined by the position of the value $(ij,k\ell)$ in the file.

A symbolic diagram of the file produced by each type of storage method will perhaps clarify the ideas involved. Trivially small blocks of four integrals are used for typographical convenience and the first two such blocks of two electron integrals are shown - the numerical values are taken from the BeH_2 example used later.

i) 1 1 1 1 0.622 2 1 1 1 0.031 2 1 2 1 0.003 2 2 2 1 0.031

2 2 2 2 0.622 3 1 1 1 0.036 3 1 2 1 0.004 3 2 1 1 0.032

The knowledge that the integrals are stored labels first then value (and the appropriate format) enables an integral to be easily identified.

ii) 0.622 0.031 0.003 0.031

0.622 0.036 0.004 0.032

Storing the integrals in order of increasing [[ij][kℓ]] and a knowledge of the ordering convention among i, j, k, ℓ enables a particular integral to be identified (e.g. (2 1, 2 1) must be the third in the list).

The relative advantages and disadvantages of these two methods are clear from the diagrams. Method (i) obviously requires more storage space than method (ii): four integers and one real number per integral compared to one real number. Method (i) has the advantage that any integrals which are zero can be omitted from the file while zero integrals must be included in method (ii) to ensure correct identification of integrals following. This latter point gives method (i) extreme flexibility since approximation methods which allow the neglect of certain small electron repulsion integrals can be conveniently handled without having to "pad out" the integral file with zeroes. Although it is not clear from the discussion so far, method (ii) is not flexible enough to enable molecular symmetry to be used to full advantage in the integral calculation. Groups of integrals which are equal by symmetry cannot be placed

```
C     CREATE CORE 'BUFFERS' TO STORE LABELS AND VALUES
      DIMENSION II(200),JJ(200),KK(200),LL(200)
      DIMENSION VALUE(200)
      NN=200
C
C     INITIALISE A COUNTER AND THE FILE MARKER
      IM=0
      IEND=0
C  ....  POSITION THE FILE
      REWIND NFILE
C  ....  SET UP LOOPS OBEYING THE ORDERING CONVENTIONS
      DO 1 I=1,N
      DO 1 J=1,I
      DO 1 K=1,I
      LTOP=K
      IF(I.EQ.K)  LTOP=J
      DO 1 L=1,LTOP
C  ....  CHECK FOR THE LAST INTEGRAL
      IF(L.EQ.N)  IEND=1
C  ....  INCREMENT THE COUNTER
      IM=IM+1
C
C  ....  COMPUTE THE CURRENT INTEGRAL (IJ,KL)
C  ....  AND STORE ITS LABELS AND VALUE
      VALUE(IM)=ERI(I,J,K,L)
      II(IM)=I
      JJ(IM)=J
      KK(IM)=K
      LL(IM)=L
C  ....  IS THE CALCULATION FINISHED
      IF(IEND.NE.0)   GO TO 2
C  ....  IS THE BUFFER FULL
      IF(IM.LT.NN)   GO TO 1
C  ....  IF IT IS, WRITE IT OUT OF FAST STORE
C  ....  AND START AGAIN
    2 WRITE(NFILE) IM,IEND,II,JJ,KK,LL,VALUE
      IM=0
    1 CONTINUE
      REWIND NFILE
C  ....  THE FILE IS NOW COMPLETE

      STOP
      END
```

```
      DIMENSION II(200),JJ(200),KK(200),LL(200),VALUE(200)
      REWIND NFILE
      SUM=0.0
    1 READ(NFILE) NN,IEND,II,JJ,KK,LL,VALUE
      DO 2 M=1,NN
      I=II(M)
      J=JJ(M)
      K=KK(M)
      L=LL(M)
      IJ=I*(I-1)/2+J
      KL=K*(K-1)/2+L
      TERM=VALUE(M)
      IF(I.EQ.J) GO TO 3
      TERM=TERM+TERM
      IF(K.EQ.L) GO TO 5
      TERM=TERM+TERM
      GO TO 5
    3 IF(K.NE.L) TERM=TERM+TERM
    5 IF(IJ.NE.KL) TERM=TERM+TERM
    2 SUM=SUM+TERM
      IF(IEND.EQ.0)   GO TO 1
      STOP
      END
```

together in the file as they are computed since the order of the integrals in the file is pre-determined by the file structure. This point will be discussed in detail in Chapter 12. The main disadvantage of method (i) - the larger storage requirement - can be attenuated by technical methods discussed in Appendix C. For maximum generality, we shall use method (i) in actual implementations.

We are now in a position to give program fragment which implements the ideas of the storage method (i). To simply show the *logic* of the process it is assumed that there is a FUNCTION available - ERI - which, given I, J, K and L, computes the integral (IJ,KL). ERI could, for example, be an implementation of the formulae of Appendix A. The variable IEND is used as a marker to signal the end of the file: IEND = 0 at the head of a block of integrals means more blocks follow, IEND ≠ 0 signals that the current block is the last.

To read and process the file created by this program we simply have to read the blocks of integrals and labels back into the first store and perform the necessary computations. The reading process is stopped by recognising the end-of-file condition IEND ≠ 0. To illustrate these ideas and to show how the permutation equivalences are used, a program is given above which reads the file and forms

$$\sum_{i,j,k,\ell=1}^{m} (ij,k\ell) \ ,$$

the sum of all m^4 integrals from the m∅ in the file.

8.7 FORMATION OF THE MATRIX $\mathsf{G(R)}$

The most common use of a file of atomic orbital electron repulsion integrals is in the formation of the matrix $\mathsf{G(R)}$ during an LCAOMO calculation. This application illustrates the method of dealing with the integrals which do not appear in the file because of the ordering convention. The formation of $\mathsf{G(R)}$ is also a good illustration of the necessity of looking at the equations of the theory from a new point of view. The definition of the elements of $\mathsf{G}(\mathsf{R})$ is

$$G_{ij} = \sum_{r,s=1}^{m} R_{rs}[2(ij,rs) - (ri,js)]$$

If this expression were to be programmed as it stands the problem posed to the programmer would be: "I am computing a particular element G_{ij}, therefore I need all integrals (ij,rs) and (ir,js)". This would require practically the whole file to be read for each element of G - a total of $\frac{1}{2}m(m+1)$ times for the whole matrix. The problem is much better posed as: "I have to hand a particular integral (ij,kℓ) and its labels i, j, k, ℓ; what are all the possible elements of G to which it contributes?" This way the file has to be read only once. Since the reading of an external file is a slow process, the first method is impossibly slow. This general conclusion holds for all applications in which a file held on slow storage must be read; steps should be taken to minimise the number of times the file is read. Table 8.1 shows the places in G where a particular integral (ij,kℓ) would contribute if the relevant elements of R are non-zero. The number of contributions depends on the possible equalities among i, j, k, ℓ. With the information of Table 8.1 we can now write a program fragment to illustrate the formation of $\mathsf{G}(\mathsf{R})$ in one "pass" of the integral file.

It is worth noting that, using the integral storage algorithm we chose in 8.6, the formation of $\mathsf{G}(\mathsf{R})$ does not depend in any way on the file being *complete* in the sense of containing m$ integrals. The program given above forms $\mathsf{G}(\mathsf{R})$ from whatever integrals occur in the file - provided the file has a marker IEND ≠ 0. Thus this G matrix method can be used for full LCAOMO calculations, for investigating the effect of neglect of certain electron repulsion integrals or for the use of semi-empirical integral values. We simply have to arrange to compute (or read) the integrals to be used, store

```
      SUBROUTINE GOFR(R,HF,N,NFILE)
C    HF  SHOULD CONTAIN THE ONE-ELECTRON
C     HAMILTONIAN ON ENTRY
      DIMENSION II(200),JJ(200),KK(200),LL(200),VALUE(200)
      DIMENSION R(10,10),HF(10,10)
      DATA COUL,EXCH/2.0,1.0/
      REWIND NFILE
    1 READ(NFILE) NN,IEND,II,JJ,KK,LL,VALUE
      DO 29 MM=1,NN
      I=II(MM)
      J=JJ(MM)
      K=KK(MM)
      L=LL(MM)
      IJ=I*(I-1)/2+J
      KL=K*(K-1)/2+L
      C1=COUL*R(I,J)*VALUE(MM)
      C2=COUL*R(K,L)*VALUE(MM)
      VAL=EXCH*VALUE(MM)
      IF(K.EQ.L) GO TO 97
      C2=C2+C2
      HF(I,K)=HF(I,K)-R(J,L)*VAL
      IF((I.NE.J).AND.(J.GE.K)) HF(J,K)=HF(J,K)-R(I,L)*VAL
   97 HF(I,L)=HF(I,L)-R(J,K)*VAL
      HF(I,J)=HF(I,J)+C2
      IF((I.NE.J).AND.(J.GE.L))  HF(J,L)=HF(J,L)-R(I,K)*VAL
      IF(IJ.EQ.KL) GO TO 29
      C3=C1
      IF(I.NE.J)  C3=C3+C1
      IF(J.GT.K) GO TO 22
      HF(K,J)=HF(K,J)-R(I,L)*VAL
      IF((I.NE.J).AND.(I.LE.K)) HF(K,I)=HF(K,I)-R(J,L)*VAL
      IF((K.NE.L).AND.(J.LE.L))  HF(L,J)=HF(L,J)-R(I,K)*VAL
   22 HF(K,L)=HF(K,L)+C3
   29 CONTINUE
      IF(IEND.EQ.0)  GO TO 1
      DO 2 J=1,N
      DO 2 I=J,N
    2 HF(J,I)=HF(I,J)
      RETURN
      END
```

Table 8.1 Contributions to $G(R)$ from permutations of the labels defining an electron repulsion integral: G is symmetric so only contributions to the upper triangle are given

Permutation	Contribution Code	Code	Position in G	Coefficient of $(ij,k\ell)$
$(ij,k\ell)$	a,b,c,d	a	i j	$2R_{k\ell}$
$(ji,k\ell)$	a,b,e,f	b	k ℓ	$2R_{ij}$
$(ij,\ell k)$	e,f	c	i k	$-R_{j\ell}$
$(ji,\ell k)$	c,d	d	j ℓ	$-R_{ik}$
$(k\ell,ij)$	a,b,p,q,	e	j k	$-R_{i\ell}$
$(\ell k,ij)$	a,b,g,h	f	i ℓ	$-R_{jk}$
$(k\ell,ji)$	g,h	g	k j	$-R_{\ell i}$
$(\ell k,ji)$	p,q	h	ℓ i	$-R_{kj}$
		p	k j	$-R_{\ell j}$
		q	ℓ j	$-R_{ki}$

* Assumed distinct: e.g. permuting i and j assumes $i \neq j$ etc.

them and the above program will form the relevant partial $G(R)$.

There are more efficient algorithms for formation of the G matrix but all of them depend on the scaling of the repulsion integrals by numerical factors and, at this stage, this process would probably obscure the logic of the method. We shall look at faster methods of forming $G(R)$ in Chapter 12.

8.8 THE USE OF AN INTEGRAL FILE IN THE VB METHOD

The implementation of the VB method using stored electron repulsion integrals in the file of 8.6 presents no new problems of principle. In using the Slater/Löwdin rules to evaluate the matrix elements organisational problems, similar to those encountered in the formation of $G(R)$, occur. However, the book-keeping problems are, in principle, very much more acute.

One integral (ij,kℓ) will occur *many times* in *many* matrix elements H_{ij}. This fact imposes rather severe sorting problems on the programmer if the file is not to be read a prohibitively large number of times. The definition and formation of $\mathbf{G}(\mathbf{R})$ is the same for all molecules and one program will work automatically for all cases, given $\mathbf{R}$ and a file of repulsion integrals. In contrast, the implementation of the VB method has to take account of the fact that the structures chosen to represent a molecular wave function will vary from molecule to molecule. A restricted form of VB method is described in Chapter 13 which describes molecules of a particular type.

8.9 APPLICATION TO BeH_2

The output file from an integral calculation program similar to the one given in the last section is listed below. The 5 AO's we have used to describe the BeH_2 molecule generate $5\not{8}$ (=120) electron repulsion integrals. It is readily seen that the integral (2 2, 1 1) is quite accurately given by $1/|R_{H1}-R_{H2}|$ and many of the many-centre repulsion integrals are small. The overlap matrix and the one-electron Hamiltonian integrals are also given for completeness, thus completing all the molecular integrals necessary for the orbital model calculation on the electronic structure of the BeH_2 molecule.

1	1	1	1	0.62203	2	1	1	1	0.03056	2	1	2	1	0.00308
2	2	1	1	0.19969	2	2	2	1	0.03056	2	2	2	2	0.62203
3	1	1	1	0.03619	3	1	2	1	0.00421	3	1	2	2	0.03175
3	1	3	1	0.00981	3	2	1	1	0.03175	3	2	2	1	0.00421
3	2	2	2	0.03619	3	2	3	1	0.00935	3	2	3	2	0.00981
3	3	1	1	0.38938	3	3	2	1	0.05008	3	3	2	2	0.38938
3	3	3	1	0.13782	3	3	3	2	0.13782	3	3	3	3	2.29207
4	1	1	1	0.27833	4	1	2	1	0.01824	4	1	2	2	0.13394
4	1	3	1	0.02345	4	1	3	2	0.02141	4	1	3	3	0.26168
4	1	4	1	0.14141	4	2	1	1	0.13394	4	2	2	1	0.01824
4	2	2	2	0.27833	4	2	3	1	0.02141	4	2	3	2	0.02345
4	2	3	3	0.26168	4	2	4	1	0.08824	4	2	4	2	0.14141
4	3	1	1	0.08479	4	3	2	1	0.01061	4	3	2	2	0.08479
4	3	3	1	0.02445	4	3	3	2	0.02445	4	3	3	3	0.35463
4	3	4	1	0.05627	4	3	4	2	0.05627	4	3	4	3	0.06248
4	4	1	1	0.32421	4	4	2	1	0.03223	4	4	2	2	0.32421
4	4	3	1	0.04162	4	4	3	2	0.04162	4	4	3	3	0.49059
4	4	4	1	0.19452	4	4	4	2	0.19452	4	4	4	3	0.10478
4	4	4	4	0.34728	5	1	1	1	-0.34225	5	1	2	1	-0.01808
5	1	2	2	-0.11022	5	1	3	1	-0.02244	5	1	3	2	-0.01893
5	1	3	3	-0.23680	5	1	4	1	-0.16120	5	1	4	2	-0.07625
5	1	4	3	-0.05162	5	1	4	4	-0.19194	5	1	5	1	0.19601
5	2	1	1	0.11022	5	2	2	1	0.01808	5	2	2	2	0.34225
5	2	3	1	0.01893	5	2	3	2	0.02244	5	2	3	3	0.23680
5	2	4	1	0.07625	5	2	4	2	0.16120	5	2	4	3	0.05162
5	2	4	4	0.19194	5	2	5	1	-0.05952	5	2	5	2	0.19601
5	3	1	1	-0.00933	5	3	2	1	-0.00000	5	3	2	2	0.00933
5	3	3	1	-0.00108	5	3	3	2	0.00108	5	3	3	3	0.00000
5	3	4	1	-0.00436	5	3	4	2	0.00436	5	3	4	3	0.00000
5	3	4	4	0.00000	5	3	5	1	0.00751	5	3	5	2	0.00751
5	3	5	3	0.00527	5	4	1	1	-0.10789	5	4	2	1	-0.00000
5	4	2	2	0.10789	5	4	3	1	-0.00229	5	4	3	2	0.00229
5	4	3	3	0.00000	5	4	4	1	-0.04296	5	4	4	2	0.04296
5	4	4	3	0.00000	5	4	4	4	0.00000	5	4	5	1	0.06864
5	4	5	2	0.06864	5	4	5	3	0.01005	5	4	5	4	0.07581
5	5	1	1	0.36767	5	5	2	1	0.03285	5	5	2	2	0.36767
5	5	3	1	0.04055	5	5	3	2	0.04055	5	5	3	3	0.46759
5	5	4	1	0.20583	5	5	4	2	0.20583	5	5	4	3	0.10131
5	5	4	4	0.34684	5	5	5	1	-0.21398	5	5	5	2	0.21398
5	5	5	3	0.00000	5	5	5	4	0.00000	5	5	5	5	0.37427

OVERLAP MATRIX

1.00000	0.08344	0.08718	0.56716	-0.58284
0.08344	1.00000	0.08718	0.56716	0.58284
0.08718	0.08718	1.00000	0.21890	0.00000
0.56716	0.56716	0.21890	0.99999	0.00000
-0.58284	0.58284	0.00000	0.00000	1.00000

ONE ELECTRON HAMILTONIAN MATRIX

-2.24290	-0.28403	-0.75084	-1.47501	1.34273
-0.28403	-2.24290	-0.75084	-1.47501	-1.34273
-0.75084	-0.75084	-8.44694	-1.86183	0.00000
-1.47501	-1.47501	-1.86183	-2.55016	0.00000
1.34273	-1.34273	0.00000	0.00000	-2.28536

SUGGESTIONS FOR FURTHER READING

"Formulas and Tables for Overlap Integrals" by R.S. Mulliken, C.A. Rieke, D. Orloff & H. Orloff in J.Chem.Phys., 17, 1248 (1949) contains expansions of the STO overlap integrals in terms of the A and B functions, plus some numerical results.

"A Study of Two-Centre Integrals Useful in Calculation on Molecular Structure I" by C.C.J. Roothaan in J.Chem.Phys., 19, 1445 (1951) discusses the STO molecular integrals which can be evaluated analytically, and gives formulae for these integrals.

"Evaluation of Molecular Integrals" by C. Guidotti and M. Maestro in La Ricerca Scientifica(Rome) 8, 1155 (1965) (in English) describes a numerical integration method for use with STO's.

"Evaluation of Multi-Centre Integrals by Polished Brute-Force Techniques" by A.C. Wahl & R.H. Land is a description of a numerical quadrature technique for STO integrals suitable for use on a large computer.

The GTF molecular integral formulae are derived in "Gaussian Expansion Methods for Molecular Integrals" by S. Huzinaga, H. Taketa & K. O-ohato in J.Phys.Soc.(Japan), 21, 2313 (1966).

"The Gaussian Function in Calculations of Statistical Mechanics and Quantum Mechanics" by I. Shavitt in Vol. 2 of "Methods in Computational Physics" (Academic Press 1963) contains derivations of some of the integral formula.

"Electronic Wave Functions I" by S.F. Boys, Proc.Roy.Soc.(London), A200, 542(1950) is the source paper and contains derivations for s-type GTF's

"Approximations for the Functions $F_m(z)$ Occurring in Molecular Calculations with a Gaussian Basis" by L.J. Shaad & G.O. Morrell in J.Chem.Phys., 54, 1965 (1971) gives methods of computing the function $F_\nu(t)$ of 8.4 rapidly.

The integral storage and retrieval method is based on the POLYATOM system, referenced in Appendix B.

"An Alternative Procedure for setting up Fock Matrices from Randomly Ordered Lists of Electron Repulsion Integrals" by A.J. Duke in Chem.Phys.Letters, 13, 76 (1971) describes a "scaling" procedure and a simple algorithm for forming the RHF matrix.

9 ORBITAL TRANSFORMATIONS

9.1 RECAPITULATION

This chapter is devoted to examinations of the effect, on a valence calculation, of changes in the orbital basis: in particular, the transformations induced by defining new AO's or MO's (or basis functions) in terms of linear combinations of the original functions. These transformations are important on chemical grounds and from a programming point of view. Typical transformations of this type are the formation of an orthogonal AO set, the computation of localised molecular orbitals and the familiar hybridisation of AO's.

The transformation properties of the sets of orbitals and the associated molecular integrals are most clearly expressed in matrix notation and so we summarise some of our earlier definitions and results in a rather more concise form. The matrix representation A of any one electron operator $\hat{A}$ in the orbital basis is given by

$$\mathsf{A} = \int dr \boldsymbol{\varphi}^\dagger \hat{A} \boldsymbol{\varphi} \qquad (9.1.1)$$

for example overlap:

$$\mathsf{S} = \int dr \boldsymbol{\varphi}^\dagger \boldsymbol{\varphi}$$

one electron Hamiltonian:

$$H = \int dr\, \boldsymbol{\varphi}^{\dagger}\, \hat{h}\, \boldsymbol{\varphi}$$

The molecular orbitals are related to the AO's by

$$\boldsymbol{\psi} = \boldsymbol{\varphi}\, \mathsf{T} \tag{9.1.2}$$

the matrix R of charges and bond orders is

$$\mathsf{R} = \mathsf{T}\,\mathsf{T}^{\dagger} \tag{9.1.3}$$

and the RHF Hamiltonian is related to R by

$$H^{F} = H + G(\mathsf{R}) \tag{9.1.4}$$

9.2 ORBITAL TRANSFORMATIONS AMONG THE AO'S

An orbital basis $\boldsymbol{\varphi}'$ can be formed by taking linear combinations of the original AO set $\boldsymbol{\varphi}$ and the relevant coefficients collected in a square matrix Y (say)

$$\boldsymbol{\varphi}' = \boldsymbol{\varphi}\, \mathsf{Y} \tag{9.2.1}$$

and, provided Y is non singular,

$$\boldsymbol{\varphi} = \boldsymbol{\varphi}'\, \mathsf{Y}^{-1} \tag{9.2.2}$$

If the matrix representation of any operator A in the basis $\boldsymbol{\varphi}'$ is given by

$$\mathsf{A}' = \int dr\, \boldsymbol{\varphi}'^{+}\, \hat{A}\, \boldsymbol{\varphi}' \tag{9.2.3}$$

then, using (9.2.1) and remembering that

$$(\mathsf{AB})^{\dagger} = \mathsf{B}^{\dagger}\, \mathsf{A}^{\dagger}$$

(for all matrices A, B), we have

$$\mathsf{A}' = \int dr\, \mathsf{Y}^{\dagger}\boldsymbol{\varphi}^{\dagger}\, \hat{A}\, \boldsymbol{\varphi}\,\mathsf{Y} = \mathsf{Y}^{\dagger}\left(\int dr\, \boldsymbol{\varphi}^{\dagger}\, \hat{A}\, \boldsymbol{\varphi}\right)\mathsf{Y}$$

that is,

$$A' = Y^{\dagger} A Y \qquad (9.2.4)$$

The effect of the transformation (9.2.1) on the R matrix can easily be found from (9.1.2) and (9.2.2):

$$\psi = \varphi T = \varphi' Y^{-1} T = \varphi' T'$$

$$\text{where } T' = Y^{-1} T$$

Now, by analogy with (9.1.3), we can define the charge and bond order matrix in the basis φ' by

$$R' = T' T'^{\dagger}$$

$$= Y^{-1} T T^{\dagger} Y^{-1\dagger}$$

that is, $$R' = Y^{-1} R Y^{-1\dagger} \qquad (9.2.5)$$

Thus the operator matrices and the charge and bond order matrices transform *contragradiently*. It is a straightforward exercise to show that the expectation value, $\langle\hat{A}\rangle$, of the operator $\hat{A}$ is given by

$$\langle\hat{A}\rangle = 2\mathrm{tr}\, R A \qquad (9.2.6)$$

But, using (9.2.4) and (9.2.5)

$$2\mathrm{tr}\, R'A' = 2\mathrm{tr}\, Y^{-1} R Y^{-1\dagger} Y^{\dagger} A Y$$

$$= 2\mathrm{tr}\, Y^{-1} R (Y Y^{-1})^{\dagger} A Y$$

$$= 2\mathrm{tr}\, Y^{-1} R A Y$$

and, since $\mathrm{tr}AB = \mathrm{tr}BA$ for any matrices A B,

$$2\mathrm{tr}\, R' A' = 2\mathrm{tr}\, R A \qquad (9.2.7)$$

The calculated value of any observable is unchanged by a linear transformation among the orbital basis functions. In particular,

$$E = 2\mathrm{tr}\, H R + \mathrm{tr}\, G(R) R$$

and (9.2.8)

$$E = 2\mathrm{tr}\, H' R' + \mathrm{tr}\, G'(R') R'$$

The LCAOMO calculation gives a total electronic energy which is unchanged by non-singular linear transformations among the basis orbitals - the self-consistent R and R' are related by (9.2.5). An orbital transformation is non-singular if it preserves the linear independence of the orbitals - crudely the new orbitals φ' do not contain repetitions. The invariance of the LCAOMO total energy with respect to linear transformations has important consequences for the implementation of the RHF method using non-orthogonal AO's.

9.3 TRANSFORMATIONS AMONG THE MO'S

By analogy with (9.2.1) a new set of molecular orbitals can be defined in terms of the original set of (9.1.2) by

$$\psi_L = \psi L \qquad (9.3.1)$$

Here, ψ_L is the new set and the transformation coefficients are collected in the matrix L. Using (9.1.2) we have

$$\psi_L = \psi L = \varphi T L = \varphi T_L \qquad (9.3.2)$$

where $T_L = T L$.

The equation

$$\psi_L = \varphi T_L$$

defines a new set of MO's in terms of the AO's and therefore a corresponding R matrix can be defined

$$R_L = T_L T_L^{\dagger} = T(L L^{\dagger})T^{\dagger}$$

using (9.3.2). If the matrix L is *unitary* - representing a transformation from an *orthogonal* set ψ to a *new orthogonal* set ψ_L - then

$$\mathsf{L}\mathsf{L}^{\dagger} = \mathsf{L}^{\dagger}\mathsf{L} = 1$$

and

$$\mathsf{R}_L = \mathsf{R} \tag{9.3.3}$$

Now, as R defines the total electron distribution we have an important theorem:

> Unitary transformations among the (occupied) MO's do not change the total LCAOMO wave function.

This result can also be obtained as an application of a well-known theorem about determinants. The LCAOMO wave function is

$$\Psi = \det\{\psi_1\bar{\psi}_1 \ldots \psi_{n/2}\bar{\psi}_{n/2}\}$$

and linear combinations of the rows or columns of a determinant do not change the value of the determinant.

The use of the symbol L in (9.3.1) and ψ_L for the transformed MO's is in anticipation of the application of (9.3.3) to form localised molecular orbitals in Chapter 13.

9.4 THE RHF EQUATIONS IN A NON-ORTHOGONAL BASIS

Referring to section 9.2 we can define an orthogonal set of AO's by a linear transformation among the original AO's:

$$\bar{\varphi} = \varphi \mathsf{X} \tag{9.4.1}$$

such that the transformed overlap matrix is unity,

$$\bar{\mathsf{S}} = \mathsf{X}^{\dagger}\mathsf{S}\mathsf{X} = 1 \tag{9.4.2}$$

The "bar" over the transformed quantities has been used in place of the prime to indicate an *orthogonal* set. Rearranging (9.4.2) the condition for the matrix X to generate an orthogonal set becomes

$$X X^{\dagger} = S^{-1} \qquad (9.4.3)$$

Thus the overlap matrix over the original AO functions must not be singular if orthogonalisation is to be successful. A discussion of numerical methods for the calculation of X is deferred until Section 9.7.

The matrix form of RHF equation, using non-orthogonal orbitals, is

$$H^F T = S T \epsilon$$

This equation is unaltered by the insertion of a unit matrix

$$H^F 1 T = S 1 T \epsilon$$

and, since $X X^{-1} = 1$ for any X,

$$H^F X X^{-1} T = S X X^{-1} T \epsilon$$

Multiplying both sides of the equation by $X^{\dagger}$ yields

$$(X^{\dagger} H^F X)(X^{-1} T) = (X^{\dagger} S X)(X^{-1} T) \epsilon \qquad (9.4.4)$$

and since

$$X^{\dagger} H^F X = \bar{H}^F$$

and

$$X^{\dagger} S X = \bar{S} = 1$$

becomes

$$\bar{H}^F \bar{T} = \bar{T} \epsilon \qquad (9.4.5)$$

In (9.4.5) we have written

$$X^{-1} T = \bar{T}$$

Equation (9.4.5) is just the RHF equation using the orthogonal basis $\bar{\varphi}$. We can now sketch the logic of the implementation of the LCAOMO method using non-orthogonal AO's, noting that

$$T = X\bar{T}$$

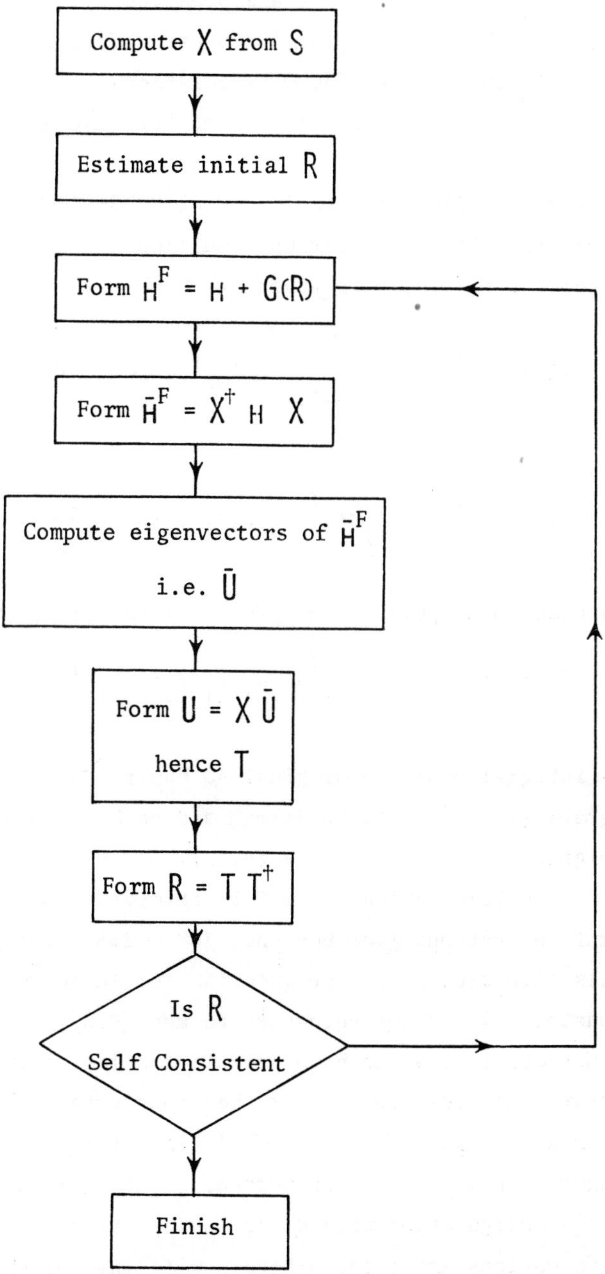

The most important computational point in this scheme is that the transformation of the electron repulsion integrals to the orthogonal basis (outlined in the next section) does not have to be performed. The simple transformation of H^F is all that is required. Thus no new procedures are required to handle non-orthogonal orbitals in LCAOMO calculations, the "overhead" due to non-orthogonality is three matrix multiplications per iteration.

9.5 TRANSFORMATION INDUCED IN THE ELECTRON REPULSION INTEGRALS

The transformation of electron repulsion integrals induced by the transformation (9.4.1) is given by

$$(\bar{i}\ \bar{j},\ \bar{k}\ \bar{\ell}) = \sum_{r,s,t,u=1}^{m} X_{ri}X_{sj}X_{tk}X_{u\ell}(rs,tu) \qquad (9.5.1)$$

where

$$(\bar{i}\ \bar{j},\ \bar{k}\ \bar{\ell}) = \int dr_1 \int dr_2 \bar{\phi}_i(r_1)\bar{\phi}_j(r_1)\frac{1}{r_{12}}\bar{\phi}_k(r_2)\bar{\phi}_\ell(r_2)$$

is an electron repulsion integral over the orthogonal basis and

$$(r\ s,\ t\ u) = \int dr_1 \int dr_2 \phi_r(r_1)\phi_s(r_1)\frac{1}{r_{12}}\phi_t(r_2)\phi_u(r_2)$$

is an integral over the original AO basis. There are m^4 new integrals $(\bar{i}\ \bar{j},\ \bar{k}\ \bar{\ell})$ to be formed and each one may involve all m^4 original integrals through (9.5.1). At first sight, it seems that the implementation of (9.5.1) involves computations involving $\sim m^8$ multiplications (for benzene, $36^8 \sim 2.8 \times 10^{12}$). Computations of this size are out of the question for large molecules. Fortunately, it is not necessary to use (9.5.1) at all in solving the RHF equations as we have seen in the last section. It *is* necessary, however, to perform the transformation (9.5.1) whenever an approximate wave function consisting of more than one orbital configuration is to be transformed. The VB method is essentially a multi-configuration method and so any attempt to perform linear transformations among the orbitals will involve this "two-electron transformation".

Examination of (9.5.1) shows that, if the equation were used as it stands, a large amount of redundant computation would be done. For example, the computation of the integrals $(\bar{2}\ \bar{2},\ \bar{1}\ 2)$ and $(\bar{2}\ \bar{2},\ \bar{1}\ \bar{1})$ *both* involve the formation of all possible products

$$X_{r2}X_{s2}(rs,tu)$$

Thus if all the products $X_{ri}X_{sj}$ and $X_{tk}X_{u\ell}(rs,tu)$ are computed and stored, the computation of the transformed electron repulsion integrals can be performed much more quickly from these products than from (9.5.1). In fact, the computation then involves $\sim m^6$ multiplications - still a formidable task. The speeding up of the computation from dependence on m^8 to m^6 is achieved at the expense of the use of large (temporary) files of stored products. Storage space for $2m^4/8$ numbers is required in addition to space for the $m^4/8$ initial integrals and the $m^4/8$ transformed integrals. It is typical for a transformation of the electron repulsion integrals to make heavier demands on computing facilities than the computation of the original AO integrals!

A program fragment which performs such a transformation on a file of electron repulsion integrals is given below. This method is thought to be the most efficient way of implementing (9.5.1).

9.6 ORBITAL TRANSFORMATIONS AND THE VB METHOD

A small addition to our treatment of the simplest VB example discussed in Section 5.3 makes it clear that, unlike in the LCAOMO method, orbital transformations have a key rôle to play in the Valence Bond theory. The Heitler-London (covalent) structure of (5.3.1) was written

$$\Psi_{HL} = N_1[\phi_A(r_1)\phi_B(r_2) + \phi_A(r_2)\phi_B(r_1)][\alpha\beta-\beta\alpha] \qquad (9.6.1)$$

If this two orbital system is transformed to an orthogonal basis

$$(\bar{\phi}_A\ \bar{\phi}_B) = (\phi_A\ \phi_B)\begin{pmatrix} X_{11} & X_{12} \\ X_{21} & X_{22} \end{pmatrix} \qquad (9.6.2)$$

```
      SUBROUTINE TWOTR(Y,M,N,NFIL1,NFIL2,NFIL3,NFIL4)
C    Y IS THE M BY N TRANSFORMATION MATRIX
C   NFIL1,NFIL4 INPUT AND OUTPUT FILES
C  NFIL2,NFIL3 ARE WORK FILES
      INTEGER R,S,RS,TU
      DIMENSION II(200),JJ(200),KK(200),LL(200),VALUE(200)
      DIMENSION Y(10,10),C(55),GAM(55)
      DATA ZERO,SMALL,NNN/0.0,1.0E-10,200/
      MT=M*(M+1)/2
      REWIND NFIL3
C   SET UP A FILE OF ALL POSSIBLE PRODUCTS OF
C  THE COLUMNS OF Y
      DO 2 I=1,N
      DO 2 J=1,I
      DO 3 R=1,M
      DO 3 S=1,R
      RS=R*(R-1)/2+S
      C(RS)=Y(R,I)*Y(S,J)
      IF(R.NE.S) C(RS)=C(RS)+Y(S,I)*Y(R,J)
    3 CONTINUE
    2 WRITE(NFIL3) C
      REWIND NFIL2
      REWIND NFIL3
C  NOW FORM FILE OF THE INTEGRAL PRODUCTS
      DO 4 I=1,N
      DO 4 J=1,I
      READ(NFIL3) C
      REWIND NFIL1
      DO 10 KT=1,MT
   10 GAM(KT)=ZERO
    5 READ(NFIL1) NN,IEND,II,JJ,KK,LL,VALUE
      DO 20 IM=1,NN
      RS=II(IM)*(II(IM)-1)/2+JJ(IM)
      TU=KK(IM)*(KK(IM)-1)/2+LL(IM)
      GAM(TU)=GAM(TU)+C(RS)*VALUE(IM)
      IF(RS.NE.TU) GAM(RS)=GAM(RS)+C(TU)*VALUE(IM)
   20 CONTINUE
      IF(IEND.EQ.0)  GO TO 5
    4 WRITE(NFIL2) GAM
      REWIND NFIL2
      REWIND NFIL4
      IEND=0
      IM=0
C   USE THE 'SCRATCH' FILES TO FORM
C  THE TRANSFORMED INTEGRALS
      DO 7 I=1,N
      DO 7 J=1,I
      REWIND NFIL3
      READ(NFIL2) GAM
      DO 7 K=1,I
      LTOP=K
      IF(I.EQ.K) LTOP=J
      DO 7 L=1,LTOP
      IF(L.EQ.N)  IEND=1
      READ(NFIL3) C
      V=ZERO
      DO 8 TU=1,MT
    8 V=V+GAM(TU)*C(TU)
      IF( ABS(V).LT.SMALL) GO TO 7
      IM=IM+1
      II(IM)=I
      JJ(IM)=J
      KK(IM)=K
      LL(IM)=L
      VALUE(IM)=V
      IF(IEND.EQ.1)  GO TO 9
      IF(IM.LT.NNN)   GO TO 7
    9 CONTINUE
      WRITE(NFIL4) IM,IEND,II,JJ,KK,LL,VALUE
      IM=0
    7 CONTINUE
      RETURN
      END
```

the covalent structure in terms of the orthogonal orbitals is

$$\Psi_{HL} = N_1[\bar{\phi}_A(r_1)\bar{\phi}_B(r_2) + \bar{\phi}_A(r_2)\bar{\phi}_B(r_1)][\alpha\beta-\beta\alpha] \qquad (9.6.3)$$

Both (9.6.2) and (9.6.3) are two-determinant functions and are *not the same* as can be verified by using (9.6.2).

The expansion of the orthogonal orbitals in terms of the non-orthogonal AO's "causes" polar structures to appear in (9.6.3), when expressed in terms of ϕ_A and ϕ_B, *with fixed weights* determined by the coefficients X_{ij} of (9.6.2). Thus a covalent structure in terms of the orthogonal orbitals does not correspond to a covalent structure in ordinary AO's.

This conclusion means that the choice of *chemical structures* in a VB calculation depends on the type of orbital basis and ordinary, intuitive, ideas of dominant structures do not hold in the orthogonal basis. Linear transformations of the orbital basis, which play a purely formal rôle in MO theory, are an important factor to be considered in VB theory: essentially an additional degree of freedom.

We have seen in 5.4 that the use of an orthogonal set of orbitals considerably simplifies the evaluation of the VB matrix elements. This simplification is, as we have seen, offset by the non-chemical nature of the structures used in describing the molecular wave function. In fact, if (9.6.3) is used for the hydrogen molecule, no chemical binding is calculated. Only when polar structures (in terms of $\bar{\phi}_A$, $\bar{\phi}_B$) are involved can a satisfactory description of the molecular electron density be achieved. This conclusion leads to an unfamiliar picture of bond formation being due to charge-transfer-like mechanisms rather than the conventional overlap based covalent bond. Computationally, the overall effect of the use of orthogonal orbitals in the VB model is to increase the *number and type of structures* which must be invoked to obtain a satisfactory picture of the bonding.

The effect of all these considerations on the computer

implementation of the VB method is to present the quantum chemist with the choice:

either (i) Use non-orthogonal AO's and program the very lengthy matrix elements (Section 5.4 or some equivalent)

or (ii) Use orthogonal orbitals with all the attendant problems of choice of structures *and* the use of the two-electron transformation (9.5.1).

Both of these options are very large computational problems and difficult to implement in general: a classic "swings and roundabouts" situation. Nothing corresponding to the automatic optimisation of the single configuration of molecular orbitals is to be found in conventional VB theory.

9.7 ORTHOGONALISATION METHODS

Although the use of an orthogonal orbital set is a computational convenience rather than a factor of chemical importance in the MO method, the choice of a *particular* set of orthogonal orbitals does have advantages in various approximations *within* the MO framework. When a "full" LCAOMO calculation is carried out the choice of orthogonal basis is immaterial since, as we have seen, the results of the calculation are invariant against the choice of basis. However, any approximation or estimation of molecular integrals, or any further model approximations (neglecting "core" electrons, say) mean that this invariance is lost and the approximate LCAOMO calculation becomes basis dependent. If we pay some attention to defining a "chemical" set of orthogonal orbitals the effect of this loss of invariance can be minimised. For example, neglect of electron repulsion integrals on differential overlap grounds is justified if these integrals *are* small - if we can choose the orbitals to make these integrals small.

Returning to equation (9.4.3) which defines the condition that a matrix X shall be an orthogonalising matrix:

$$\mathsf{X}^{\dagger} \mathsf{X} = \mathsf{S}^{-1} \tag{9.7.1}$$

we discuss various ways of finding solutions X of this equation.

First, it is clear that any solution of (9.7.1) is determined only to within a unitary transformation as

$$\mathbf{X\,B\,B^{\dagger}\,X^{\dagger}} = (\mathbf{X\,B})(\mathbf{X\,B})^{\dagger} = \mathbf{S}^{-1}$$

provided $\mathbf{B}$ is unitary ($\mathbf{B\,B^{\dagger}} = \mathbf{1}$). The simplest case is given by assuming $\mathbf{B} = \mathbf{1}$. If the matrix $\mathbf{X}$ is constrained to be symmetric then $\mathbf{X}^{\dagger} = \mathbf{X}$ and the solution of (9.7.1) is simply

$$\mathbf{X} = \mathbf{S}^{-\frac{1}{2}} \qquad (9.7.2)$$

The use of a symmetrical orthogonalisation matrix is said to *symmetrically orthogonalise* the atomic orbitals. It is possible to prove that the symmetrically orthogonalised set of orbitals is the set of orbitals which is closest to the original set consistent with orthogonality. This orthogonalisation method is therefore widely used since the orthogonalised orbitals can still, with some justification, be called atomic orbitals. Also important from a molecular point of view is the fact that the symmetrically orthogonalised AO's still retain the molecular symmetry of the original AO's. A symmetry operation inducing a given linear combination among the AO's induces the *same* linear combination among the orthogonal AO's. In our BeH_2 example the two hydrogen 1s AO's are exchanged by reflection in a plane through the central Be atom and so are the symmetrically orthogonalised 1s AO's. This last idea, that we can continue to speak of the orthogonal set in AO nomenclature, is a consequence of the form of the orthogonal orbitals: each $\bar{\phi}_i$ is "mainly" ϕ_i with small "correction terms" in the region of each overlapping AO, provided that no overlap integrals are very large. Fig. 9.1 shows the qualitative forms of AO's of BeH_2 after symmetrical orthogonalisation.

The computation of $\mathbf{S}^{-\frac{1}{2}}$ is quite straightforward and uses a matrix diagonalisation technique. If the (unitary) matrix which reduces $\mathbf{S}$ to the diagonal form $\mathbf{D}$ is $\mathbf{E}$, then

$$\mathbf{E^{\dagger}\,S\,E} = \mathbf{D}$$

and

$$S = E D E^{\dagger} \qquad (9.7.3)$$

since

$$E E^{\dagger} = 1. \qquad (9.7.4)$$

Any power of S is therefore given by the corresponding power of the right hand side of (9.7.3), for example

$$S^2 = (E D E^{\dagger})^2 = E D E^{\dagger} E D E^{\dagger}$$

$$= E D^2 E^{\dagger}$$

using (9.7.4). In general

$$S^n = E D^n E^{\dagger}$$

and therefore any polynomial, and by implication any analytic function, of S can be expanded in powers of D, the advantage being that functions of D are simply the corresponding functions of the diagonal elements. The polynomial expansion of the function $x^{-\frac{1}{2}}$ therefore ensures that

$$S^{-\frac{1}{2}} = E D^{-\frac{1}{2}} E^{\dagger} \qquad (9.7.5)$$

where

$$D^{-\frac{1}{2}} = \begin{pmatrix} D_{11}^{-\frac{1}{2}} & 0 & 0 \\ 0 & D_{22}^{-\frac{1}{2}} \cdots\cdots & 0 \\ 0 \cdots\cdots & \cdots & \cdots \\ 0 & 0 \cdots\cdots & D_{nn}^{-\frac{1}{2}} \end{pmatrix}$$

Diagonalisation of S and the transformation (9.7.5) are the only numerical techniques required.

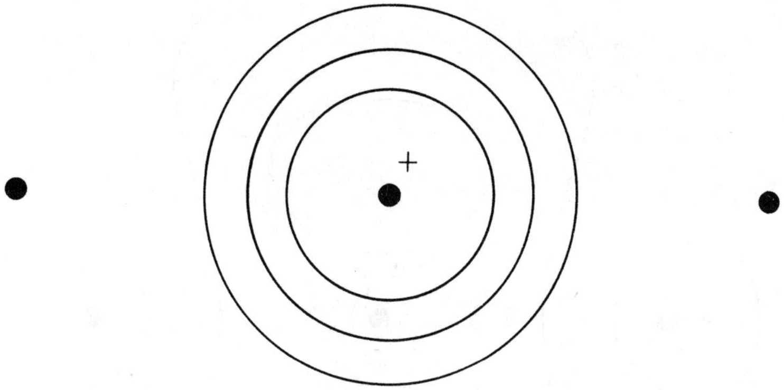

Figure 9.1a   Orthogonalised beryllium 1s orbital

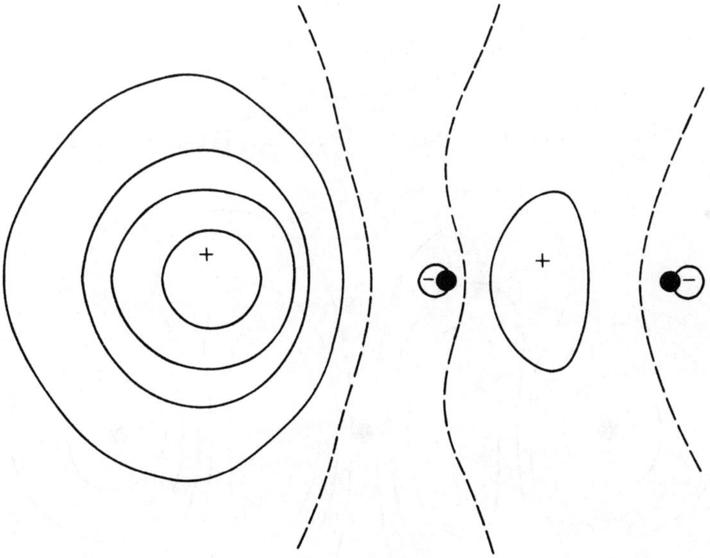

Figure 9.1b   Orthogonalised hydrogen 1s orbital

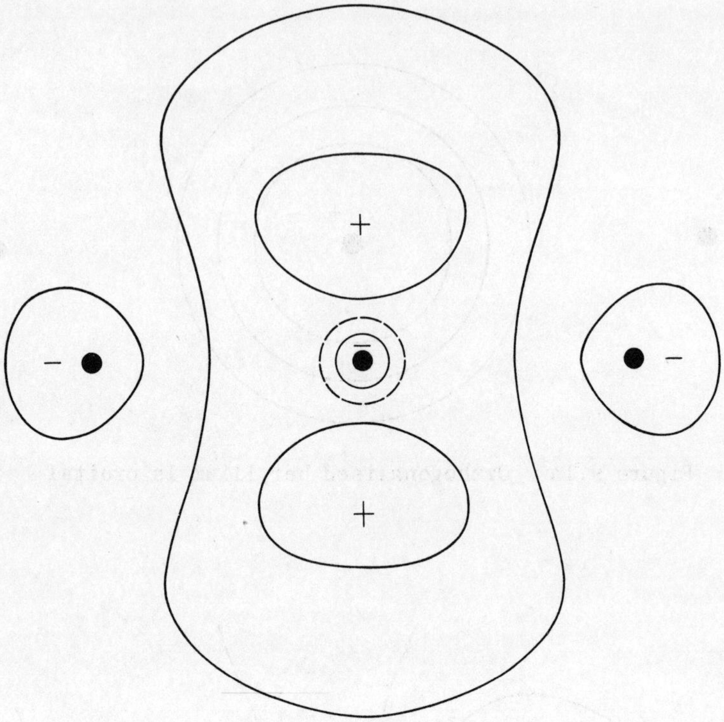

Figure 9.1c Orthogonalised beryllium 2s orbital

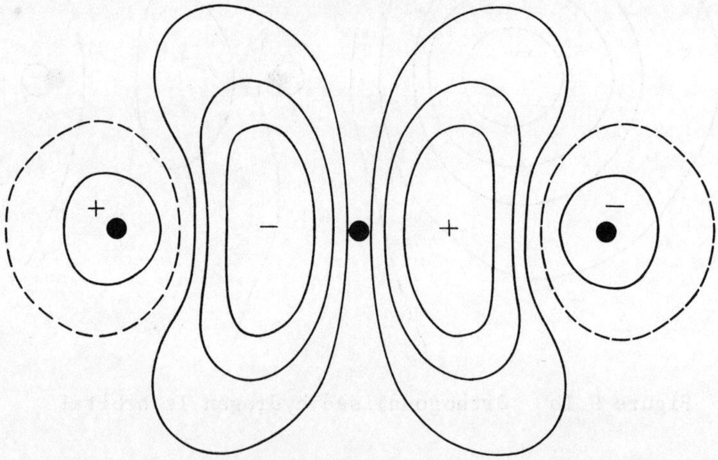

Figure 9.1d Orthogonalised beryllium 2p orbital

Another widely used orthogonalisation method (particularly in atomic work where symmetry problems are more easily solved) is based on the idea of building up a set of orthogonal orbitals by forcing successive members to be orthogonal to the existing functions. This is the Schmidt method and is well known as a theorem in linear algebra. The basic formula is the one for two orbitals: if ϕ_2 is to be made orthogonal to ϕ_1 *without changing* ϕ_1 then the necessary transformation is:

$$\bar{\phi}_1 = \phi_1$$
$$\bar{\phi}_2 = \phi_2 - S_{12}\phi_1 \tag{9.7.6}$$

where

$$S_{12} = \int dr\phi_1(r)\phi_2(r)$$

It is convenient to renormalise $\bar{\phi}_2$ by the multiplicative factor $(1-S_{12}^2)^{-\frac{1}{2}}$. The obvious extension of (9.7.6) is a recessive formula

$$\bar{\phi}_{i+1} = \phi_{i+1} - \sum_{j=1}^{i} \bar{\phi}_j S_{ij} \tag{9.7.7}$$

When this process is carried through, and the coefficients collected, the orthogonalising matrix χ for Schmidt orthogonalisation has characteristic "upper triangular" form corresponding to each new orbital being a linear combination of the earlier ones.

$$\chi = \begin{pmatrix} X_{11} & X_{12} & X_{13} & \cdots & X_{1n} \\ 0 & X_{22} & X_{23} & \cdots & X_{2n} \\ 0 & 0 & X_{33} & \cdots & X_{3n} \\ 0 & 0 & 0 & \cdots & \cdot \\ \cdot & \cdot & \cdot & \cdots & \cdot \\ 0 & 0 & 0 & \cdots & X_{nn} \end{pmatrix}$$

The calculation of a matrix of this triangular form is typical of matrix inversion procedures and, of course, the matrix X is related to S^{-1} by (9.7.1). The computational implementation of the method closely parallels matrix inversion methods and also suffers from the same difficulties with cumulative errors. Generation of the matrix X using (9.7.7) causes the building up of rounding errors particularly if some off diagonal elements of S are large (S is almost singular). Numerical methods for the Schmidt procedure have been developed which avoid the build-up of errors, and we give a sketch of one of these (due to P.-O. Löwdin) now.

Re-writing (9.7.7) in a more compact form in terms of the overlap integrals over the original AO's we have

$$\bar{\phi}_{i+1} = \phi_{i+1} - \boldsymbol{\varphi}_i \mathsf{S}_i^{-1} s_i \qquad (9.7.8)$$

where $\boldsymbol{\varphi}_i$ is the row matrix of the first i AO's

$$\boldsymbol{\varphi}_i = (\phi_1, \phi_2, \ldots \phi_i)$$

S_i is the overlap matrix generated by these functions

$$\mathsf{S}_i = \int dr \boldsymbol{\varphi}_i \boldsymbol{\varphi}_i^{\dagger} \qquad (9.7.9)$$

and s_i is the column of overlap integrals between

$$\phi_{i+1} \text{ and } \boldsymbol{\varphi}_i$$

$$s_i = \int dr\, \phi_{i+1} \boldsymbol{\varphi}_i^{\dagger} \qquad (9.7.10)$$

Thus, in order to compute the orthogonalisation coefficients, we must invert S_i for i=1, 2, ... n-1. Now any matrix can be partitioned into sub-matrices of smaller dimension, and in particular the addition of one more row and column to S_i to form S_{i+1} involves s_i.

$$\mathsf{S}_{i+1} = \begin{pmatrix} \mathsf{S}_i & s_i \\ s_i^{\dagger} & 1 \end{pmatrix} \qquad (9.7.11)$$

and it is easy to verify (by multiplication of (9.7.11) and (9.7.12)) that

$$\mathsf{S}_{i+1}^{-1} = \begin{pmatrix} \mathsf{S}_i^{-1} + \mathsf{S}_i^{-1}\, s\, s_i^{\dagger}\, \mathsf{S}_i^{-1}/D_i & -\mathsf{S}_i^{-1}\, s_i/D_i \\ -\, s_i^{\dagger}\, \mathsf{S}_i^{-1}/D_i & 1/D_i \end{pmatrix} \qquad (9.7.12)$$

where D_i, a scalar, is $1 - s_i^{\dagger}\, \mathsf{S}_i^{-1} s_i$. Equation (9.7.12) ensures that each S_{i+1} is easily inverted if the inverse of previous member of the series, S_i^{-1}, is known. Further, comparing the last column of (9.7.12) with (9.7.8), it is clear that the orthogonalisation coefficients are contained in this column - apart from normalisation. Squaring (9.7.8) and integrating, shows that the normalisation constant is $D_i^{\frac{1}{2}}$ so that this "successive inversion" of the AO overlap matrix generates the Schmidt orthogonalisation coefficients (and normalising factors). The process is started with i=1,

$$\mathsf{S}_1 = \mathsf{S}_1^{-1} = 1$$

$$s_1 = S_{12} \; ; \quad D_1 = 1 - S_{12}^2$$

(giving the result of (9.7.6)) and continued until the whole AO set has been orthogonalised.

In addition to the numerical difficulties associated with the Schmidt method, there are difficulties of interpretation involved in using the orthogonal set produced. The set $\bar{\boldsymbol{\varphi}}$ of orthogonal orbitals depends on the *ordering* of the members of $\boldsymbol{\varphi}$. The symmetry properties of $\boldsymbol{\varphi}$ are destroyed in forming $\bar{\boldsymbol{\varphi}}$ and the orthogonal orbitals cannot be identified as "corrected" AO's. In the case of BeH_2, if we take one of the hydrogen AO's as starting point, there are 24 possibilities for the order of the remaining four orbitals. In every case the other hydrogen AO is transformed into a linear combination of the remaining AO's: a combination of at least two and at most five AO's. Thus the reflection symmetry of the hydrogen AO's is destroyed by the Schmidt orthogonalisation.

For some purposes the fact that the Schmidt method can leave orbitals unchanged is an advantage. It is usual to assume that the bonding among first row atoms of the periodic table involves the 2s and 2p orbitals but not the 1s shell. If the orbitals of a first row atom are symmetrically orthogonalised then the 1s orbital is forced to change because of its overlap with the 2s orbital. Using the Schmidt method the 1s orbitals on each atom can be left unchanged and still be orthogonal to the "valence orbitals".

The Schmidt method is not quite adequate to deal with the "core-valence" separation in the general molecular situation since it is usually desirable to leave *several* inner orbitals uncontaminated by valence orbital components *and* to generate an orthogonal set. With the Schmidt method in mind we can easily generate such a partial orthogonalisation procedure. We refer to a set of n_c "core" orbitals ϕ_{i_c} and a set of n_v "valence" orbitals ϕ_{i_v} which are non-orthogonal, and require a linear transformation to a new set of valence orbitals $\bar{\phi}_{i_v}$ which are orthogonal to all the core orbitals ϕ_{i_c},

$$\int dr\ \phi_{i_c}(r)\bar{\phi}_{j_v}(r) = 0 \tag{9.7.13}$$

We write each new valence orbital as a linear combination of the old valence orbital and "correction" terms in the region of each core orbital,

$$\bar{\phi}_{j_v} = \phi_{j_v} + \sum_{i_c=1}^{n_c} W_{i_c j_v}\ \phi_{i_c}$$

and insist that the elements of the matrix $\mathbb{W}$ shall be chosen to make each orthogonality condition (9.7.13) hold

$$\int dr\phi_{k_c}(r)\bar{\phi}_{j_v}(r) = \int dr\phi_{k_c}(r)\phi_{j_v}(r) + \sum_{i_c=1}^{n_c} W_{i_c j_v}$$

$$\times \int dr\phi_{k_c}(r)\phi_{i_c}(r) = 0$$

Collecting these equations we have

$$S_c W + S_{cv} = 0$$

where

$$(S_c)_{i_c j_c} = \int dr\phi_{i_c}(r)\phi_{j_c}(r)$$

$$(S_{cv})_{i_c j_v} = \int dr\phi_{i_c}(r)\phi_{j_v}(r)$$

that is

$$W = - S_c^{-1} S_{cv} \qquad (9.7.14)$$

The core overlap matrix is always non-singular and so the required coefficients are contained in the matrix W. This orthogonalisation procedure is a generalisation of the Schmidt method in the sense of producing orthogonality between two *sets* of orbitals leaving one set unchanged.

9.8 COMPOSITE ORTHOGONALISATION METHODS

It is useful for interpretive purposes to combine the advantages of the symmetrical and Schmidt orthogonalisation methods in defining a *localised orthogonal atomic orbital set*. We can achieve this by performing a series of transformations on the "raw" AO's:

i) Orthogonalise the valence orbitals against the core orbitals by the last method of 9.7;

ii) Hybridise the resulting valence AO's in a way consistent with the conventional chemical bonding scheme: hybrids pointing along each bond;

iii) Dispose of all remaining non-orthogonalities by symmetrical orthogonalisation.

Thus, in matrix language, the steps are:

i) $$\boldsymbol{\varphi}_W = \boldsymbol{\varphi}\, \mathsf{W} \tag{9.8.1}$$

ii) $$\boldsymbol{\varphi}_H = \boldsymbol{\varphi}_W \mathsf{V} = \boldsymbol{\varphi}\, \mathsf{W}\, \mathsf{V} \tag{9.8.2}$$

where V is a unitary matrix whose columns define the hybrid AO's

iii) $$\bar{\boldsymbol{\varphi}} = \boldsymbol{\varphi}_H \mathsf{S}_H^{-\frac{1}{2}} \tag{9.8.3}$$

where

$$\mathsf{S}_H = \int dr\, \boldsymbol{\varphi}_H^{\dagger} \boldsymbol{\varphi}_H = \mathsf{V}^{\dagger}\, \mathsf{W}^{\dagger}\, \mathsf{S}\, \mathsf{W}\, \mathsf{V} \tag{9.8.4}$$

is the overlap matrix over the hybrid AO's.

Collecting these equations and dropping the temporary subscripts gives

$$\bar{\boldsymbol{\varphi}} = \boldsymbol{\varphi}\, \mathsf{W}\, \mathsf{V}\, \mathsf{S}_H^{-\frac{1}{2}} \tag{9.8.5}$$

as a localised orthogonal AO basis.

Again, it must be emphasised that this orthogonalisation method has no advantage over the simple $\mathsf{S}^{-\frac{1}{2}}$ in the solution of the RHF equation as the results are invariant. This form of the AO basis is, however, clearly adapted to a model treatment of the inner shells of a molecule. It can also be shown that $\bar{\boldsymbol{\varphi}}$ defined by (9.8.5) is an excellent basis for use of differential overlap methods. If the R matrix is transformed to the basis (9.8.5) the "bond densities" are shown in the most convenient way. Figure 9.2 shows the spatial form of the orthogonal sp hybrid formed using (9.8.5).

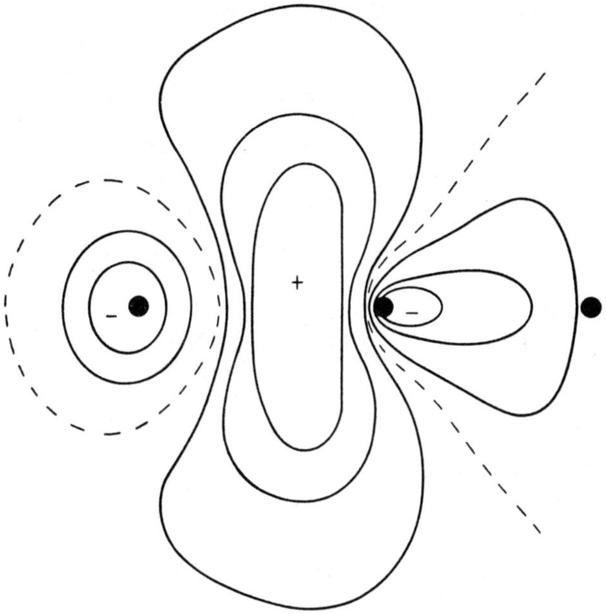

Figure 9.2 The orthogonalised beryllium sp hybrid

9.9 APPLICATION TO BeH_2

The most important result obtained in this chapter is the method of solution of the RHF equation for a non-orthogonal basis. We have already computed the AO basis molecular integrals in Chapter 8. If we implement the scheme flow-charted in Section 9.4 we can obtain the LCAOSCFMO's for BeH_2 ground state. Using the programs listed in Appendix C we obtain the self-consistent orbitals ($\mathbb{T}$), orbital energies (ε_i) and charge and bond order matrix ($\mathbb{R}$) listed in Table 9.1 Figure 9.3 gives contour diagrams of the three occupied molecular orbitals - we defer any discussion of the forms of these orbitals and the elements of the charge and bond order matrix until Chapter 10.

Table 9.1 MO Results for BeH_2

ε_i	-4.5092	-0.5024	-0.4625
T	-0.0061	0.4540	0.4266
	-0.0061	0.4540	-0.4266
	0.9892	-0.2034	0.0
	0.0501	0.4048	0.0
	0.0	0.0	-0.4593

R	0.3881	0.0238	-0.0983	0.1834	-0.1956
		0.3881	-0.0983	0.1834	0.1956
			1.0199	-0.0328	0.0
				0.1664	0.0
					0.2100

Total Electronic Energy = -18.46311 a.u.
Nuclear Repulsion Energy = 3.4 a.u.
Total Energy = -15.06311 a.u.

Atom energies Be -13.93906, H -0.486 a.u.
(using same GTF expanded AO's)

Binding Energy = 0.152 a.u. (∿200 kJ mol^{-1} per Be-H bond)

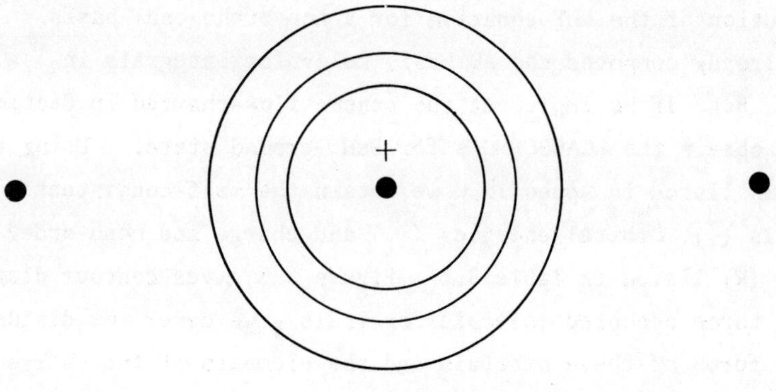

Figure 9.3a Molecular Orbital 1 for BeH_2

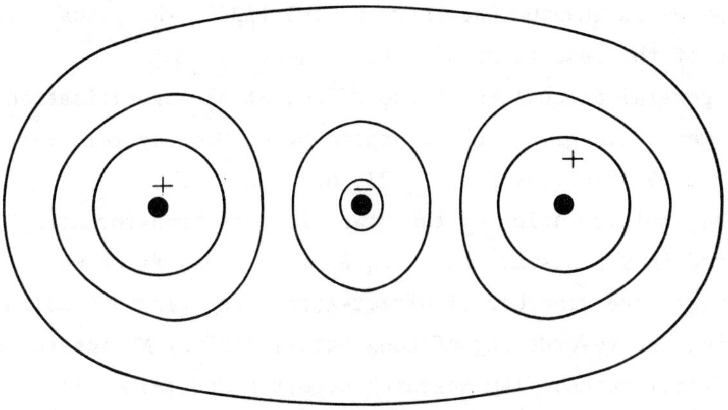

Figure 9.3b Molecular Orbital 2 for BeH_2

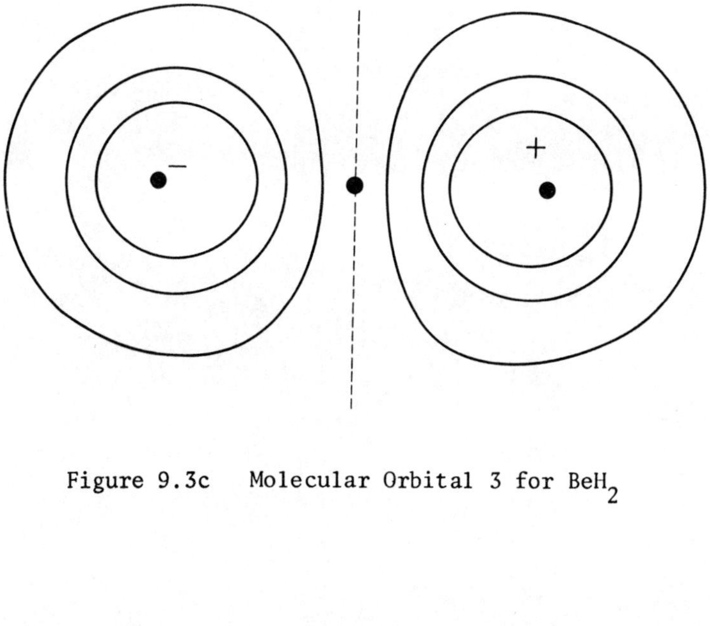

Figure 9.3c Molecular Orbital 3 for BeH_2

SUGGESTIONS FOR FURTHER READING

The treatment of orthogonalisation methods parallels the discussion in "On the non-orthogonality Problem" by P.-O. Löwdin in Advances in Quantum Chemistry 5, 185 (1970) who gives proofs of some of the results simply stated here.

A general discussion of the effect of orthogonalisation on the VB method is given in the papers by R. McWeeny referenced in Chapter 5 (Proc.Roy.Soc., A223, 63, 306 (1954))

The implementation of the "two electron transformation" described in 9.5 assumed only sequential-access files were available: see "The Use of Direct-Access devices in Problems requiring the re-ordering of Long Data-lists" by M. Yoshimine for a faster method (IBM Research Report RJ555 (#11634): available on request from IBM Research Laboratory, San Jose, California 95114, U.S.A.)

10 POPULATION ANALYSIS AND PHYSICAL INTERPRETATION

10.1 QUALITATIVE AND QUANTITATIVE INFORMATION

In Chapter 1 we saw that two types of information should come out of an approximate solution of the Schrödinger equation for a molecular system. The "chemical" information, although a quantitative estimate of the electron density changes on bonding, is used in rather qualitative ways; the full electron density function is not used but questions like "how many electrons move into the region between nuclei α and β?" occur. Questions of a more precise nature are asked of the calculation by spectroscopists and chemical physicists: "what is the dipole moment of the molecule?", "what is the ionisation potential or first excitation energy of the system?" etc. These two separate areas of interest are, of course, related and might be classified as the qualitative and quantitative interpretation of the density function (2.6.5) respectively.

The orbital model we have used to obtain approximate solutions of the Schrödinger equation throws much of the analysis of the density function into the interpretation of the charge and bond-order matrix R. To avoid this occurrence of factors of 2 throughout the equations we will use the matrix $\mathsf{P} = 2\mathsf{R}$. The analysis is biased towards the use of the MO charge and bond-order matrix

$$\mathsf{P} = 2\mathsf{R} = 2\mathsf{T}\,\mathsf{T}^{\dagger} \qquad (10.1.1)$$

but any approximate orbital-model wave function yields a P matrix through the use of (2.6.5).

10.2 POPULATION ANALYSIS

Perhaps the most important point to be stressed before beginning an interpretation of the charge and bond-order matrix is that the approximate molecular wave function and the corresponding one-electron density function are continuous functions of space. The charge and bond-order matrix is "only" a collection of coefficients multiplying spatial functions and the over-all distribution is best seen as a contour diagram if ambiguities of interpretation arise.

With this reservation in mind we re-capitulate the result of Chapter 5 relating the density function P(r) to the charge and bond-order matrix P

$$P(r) = \boldsymbol{\varphi}(r)\,\mathsf{P}\boldsymbol{\varphi}^{\dagger}(r) = \sum_{i,j=1}^{m} \phi_i(r)\phi_j(r)P_{ij} \qquad (10.2.1)$$

Thus the elements P_{ij} are the coefficients of the spatial products $\phi_i\phi_j$ and measure the *weight* with which a given orbital product contributes to the total electron density. It is, therefore, a priori possible for a product which is vanishingly small over all space to have a large coefficient leading to a nonsensical interpretation of the relative magnitudes of the P_{ij}. This situation does not often occur - the variational solution of the Schrödinger equation will "try" to put electrons where there is most energy to be had and therefore avoid occupying spatial distributions which are very small. Nevertheless, in comparing elements of P involving electrons in orbitals of very different energies care must always be taken to ensure that like quantities are being compared (e.g. σ and π electrons in aromatics). So much for precautions, we can now look at the analysis.

Particular importance is attached to the elements of P along the principal diagonal - the P_{ii} - since these are the contributions to the total electron density of the squares of the AO's - the *occupation numbers* of the orbitals and therefore (when integrated) the *charge* in each orbital

$$\int dr\; P_{ii}\phi_i^{\,2}(r) \;=\; P_{ii} \tag{10.2.2}$$

The interpretation of the off-diagonal elements has a similar pictorial meaning as "the number of electrons in the overlap region $\phi_i\phi_j$". This interpretation is particularly useful for formally bonded or near-neighbour orbitals. The interpretation of bond orders between distant orbitals (which are usually small) is not of very great value. This interpretation of the elements of P is strictly only true if the AO's are orthogonal since only then do the orbital "charges" add up to the total number of electrons in the molecule,

$$\mathrm{tr}\,\mathsf{P}\,\mathsf{S} \;=\; n$$

and if $\quad \mathsf{S} = \mathbf{1} \qquad \mathrm{tr}\,\mathsf{P} = n.$

It is possible to "re-normalise" the orbital-product distributions, dividing each one by the corresponding overlap term, $\phi_i\phi_j/S_{ij}$, and re-defining the charges and bond orders as $p_{ij} = P_{ij}S_{ij}$. This definition has some formal advantages but is also thrown into disarray by small overlap terms.

There are two obvious limitations on the utility of our analysis so far.

i) The MO calculation is invariant against linear transformations among the AO functions but the charge and bond-order matrix when *expressed in terms of the transformed AO's* (e.g. hybrids) is often very different from the original matrix although both give the same density function

ii) If the AO basis is extended to include "excited" orbitals (or if the basis functions χ_i are used in place of the AO's) then a much larger P matrix is obtained and the chemical interpretation of the elements is difficult.

Both of these difficulties can be overcome by using a coarser breakdown of the electron density. If we confine attention to the electronic populations of *atoms* and *inter-atomic* regions then the dependence of the P matrix on size and nature of orbital basis is minimised.

Re-writing (10.2.1) in terms of the p_{ij} and the renormalised AO products, we have

$$P(r) = \sum_{i,j=1}^{m} p_{ij}(\phi_i\phi_j/S_{ij})$$

which breaks down into atomic and inter-atomic terms as follows:

$$P(r) = \sum_{\alpha=1}^{N} \{ \sum_{i,j\varepsilon\alpha} p_{ij}(\phi_i\phi_j/S_{ij})\}$$

$$+ \sum_{\alpha\neq\beta=1}^{N} \{ \sum_{i\varepsilon\alpha} \sum_{j\varepsilon\beta} p_{ij}(\phi_i\phi_j/S_{ij})\} \qquad (10.2.3)$$

where the notation $i\varepsilon\alpha$ means the summation is to run over orbitals ϕ_i which are centred on atom α and similarly for $j\varepsilon\beta$. Clearly (10.2.3) can be written

$$P(r) = \sum_{\alpha=1}^{N} P_\alpha(r) + \sum_{\alpha\neq\beta=1}^{N} P_{\alpha\beta}(r) \qquad (10.2.4)$$

emphasising the atomic and inter-atomic contributions to the electron density. Performing a similar summation over the p_{ij} elements *alone* gives a coarse discription of the partitioning of the electrons among the atoms and inter-atomic regions. We can, for example, with some physical justification say that

$$\sum_{i,j\varepsilon\alpha} p_{ij} \qquad (10.2.5)$$

is, in some sense, the number of electrons associated with atom α and therefore the overall charge on this atom in the molecule is

$$Q_\alpha = \sum_{i,j\varepsilon\alpha} P_{ij} - Z_\alpha \tag{10.2.6}$$

These definitions although convenient and physically transparent leave us with an awkward "book-keeping of charge" problem: clearly, if the molecule carries a net charge q,

$$\sum_\alpha Q_\alpha \neq q \tag{10.2.7}$$

by comparison of (10.2.5) with the full expression (10.2.3) - the inter-atomic terms have been omitted. If we wish to divide the electron density up in such a way that (10.2.7) can be satisfied, we must divide up the charge associated with the inter-atomic regions among the atoms themselves. Since this last task involves forcing a rather artificial interpretation on the electron density, it is not surprising that there is no physically satisfactory way of carrying it out. There are electrons *between* the atoms and they can only be *formally* associated with individual atoms. Mulliken has suggested that the electron density $P_{\alpha\beta}(r)$ be *equally* partitioned among α and β and so, defining

$$Q_\alpha = \sum_{i,j\varepsilon\alpha} P_{ij} + \tfrac{1}{2}\sum_{j\varepsilon\beta}\sum_{i\varepsilon\alpha} P_{ij} \tag{10.2.8}$$

satisfies (10.2.7).

The summation over all the orbitals on a given atom ensure that (10.2.8) is only slightly dependent on the number of such orbitals and the invariance conditions discussed in Chapter 9 mean that local transformations of the AO's of each atom (hybridisation) do not change Q_α.

10.3 POPULATION ANALYSIS IN PRACTICE

The charge and bond-order matrix obtained from a LCAOMO calculation is by far the most common and there are a few points which can be made about this special case. There are cases when the $\mathbf{R}$ matrix is completely determined by symmetry - the π system of ethylene or the H_2 molecule - when a small AO basis is used. If we take specifically a homopolar two electron bond formed by the combination of two AO's in the LCAOMO method, the occupied MO coefficients are necessarily $1/\sqrt{2}$ (if the orbitals are orthogonal) and so

$$\mathbf{R} = \begin{pmatrix} 0.5 & 0.5 \\ 0.5 & 0.5 \end{pmatrix}$$

independent of the AO's. Thus there are no degrees of freedom and the variation principle cannot act. Use of the VB method with the same orbital basis gives a more flexible function and therefore a more realistic matrix. For ethylene, the π bond order changes from 0.5 to 0.34 on using a more flexible form of function. A more spectacular example is provided by the π system of benzene (using 6 $2p_\pi$ AO's) where all the 36 elements of $\mathbf{U}$ are determined by symmetry. These very extreme examples show that there are fewer degrees of freedom in the MO charge and bond-order matrix than would appear at first sight (the essential requirement for a *variational* solution is the existence of more than one symmetry orbital of each species - see Chapter 12).

Symmetry sometimes restricts the freedom of action of the variation principle but in all cases the MO coefficient matrix must be unitary (for orthogonal orbitals) and this fact always reduces the number of independent R_{ij} elements. In the case of a *polar* two-orbital bond the MO coefficients are now variationally determined:

$$\mathbf{U} = \begin{pmatrix} c & -s \\ s & c \end{pmatrix}$$

but the *bond order* is determined by the *bond polarity*

$$R = \begin{pmatrix} c^2 & 2sc \\ 2sc & s^2 \end{pmatrix}$$

and so there is only one degree of freedom here.

The transformation of the AO basis to one which gives a population analysis in the most chemically meaningful form is discussed in detail in Chapter 13.

A completely different problem associated with the attempt to summarise a continuous charge distribution in the form of a few populations occurs when "diffuse" orbitals are used - orbitals which are large compared to the accepted covalent radius of the atom. The 4s orbital of the first transition series is a familiar example. In manganese this orbital is "larger" than the whole of the MnO_4^- ion and it is difficult to justify the occupation number of this orbital contributing to the charge on the manganese atom.

These remarks on the limitations of the discrete analysis of the electron density should not be taken to mean that the whole analysis is worthless; on the contrary it has been found to be a very worthwhile way of summarising the mass of information in the molecular wave function and of correlating computational results with empirical chemistry.

10.4 COMPUTATION OF MOLECULAR PROPERTIES

A full treatment of the computation of observable molecular properties (other than orbital energies) demands the use of perturbation theory. The method of detection of dipole moments, polarisabilities, NMR chemical shifts, etc., corresponds to adding (ideally) small terms to the molecular Hamiltonian and computing the way in which the molecular electron density responds to such perturbations. However, it is possible, computationally and conceptually, to regard the so-called first-order properties of the molecule as existing in their own right independent of the measuring process. The molecular dipole moment and polarisability are both "measured" by using an applied electric field but we normally say the unperturbed molecule "has a dipole moment". The calculation of these first-order properties simply involves the computed density functions and certain molecular integrals involving the relevant molecular operator. If the operator is simply a sum of one-electron operators then the matrix $\mathbf{O}$, with elements

$$O_{ij} = \int dr_1 \phi_i(r_1)\hat{O}(1)\phi_j(r_1) \tag{10.4.1}$$

is all that is required, since the mean value of

$$\sum_{i=1}^{n} \hat{O}(i)$$

is simply

$$\langle\hat{O}\rangle = \mathrm{tr}\,\mathbf{P}\,\mathbf{O} \tag{10.4.2}$$

The most common one-electron properties are the multiple moments of the electron distribution, here the operator is

$$\hat{O}(i) = x_i^{\ell} y_i^{m} z_i^{n}$$

and (10.4.2) evaluated for this operator and added to the corresponding nuclear term

$$\sum_{\alpha=1}^{N} X_\alpha^{\ell}\, Y_\alpha^{m}\, Z_\alpha^{n}$$

gives the molecular multipole moment.

The situation is a little more involved when two-electron operators occur in the perturbing term. In these cases the AO representation of the electron-pair density function of (2.6.2) is needed: the coefficients $P_2(i,j,k,\ell)$ in

$$P_2(r_1 r_2) = \sum_{i,j,k,\ell=1}^{m} \phi_i(r_1)\phi_j(r_2)P_2(i,j,k,\ell)\phi_k(r_1)\phi_\ell(r_2) \tag{10.4.3}$$

These coefficients, together with the molecular integrals

$$O'_{i,j,k,\ell} = \int dr_1 \int dr_2 \phi_i(r_1)\phi_j(r_2)\hat{O}'(1,2)\phi_k(r_1)\phi_\ell(r_2) \qquad (10.4.4)$$

(where $\hat{O}'(i,j)$ is a typical two-electron operator), define the mean value of

$$\sum_{i>j=1}^{n} \hat{O}'(i,j)$$

In the particular case of the single determinant LCAOMO wave function the coefficients $P_2(i,j,k,\ell)$ are determined by the elements of P; the "correlated" motion of electron pairs is determined by the charge density and the Pauli principle, in fact

$$P_2(i,j,k,\ell) = P_{ik}\,P_{j\ell} - \tfrac{1}{2}P_{i\ell}\,P_{jk} \qquad (10.4.5)$$

and hence

$$\langle\hat{O}'\rangle = \tfrac{1}{2} \sum_{i,j,k,\ell=1}^{m} P_2(i,j,k,\ell)O'_{i,j,k,\ell} \qquad (10.4.6)$$

The development of the perturbation theories necessary to evaluate molecular properties which depend on *changes* in electron density brought about by a perturbing operator is outside the scope of our computational treatment and some suitable references are given in the reading list. The two general approaches to the effect of a perturbation parallel the two main orbital models: "MO" and "VB". The "Hartree-Fock" or "self-consistent" perturbation theory retains the single determinant form of the wave function and seeks corrections to the orbitals in the presence of the perturbation. The "Rayleigh-Schrödinger" perturbation theory seeks corrections to the total molecular wave function by adding in "excited state" determinants, giving a multi-configuration wave function in the presence of the perturbing operator.

SUGGESTIONS FOR FURTHER READING

Systematic population analysis is usually associated with R.S. Mulliken's paper "Electronic Population Analysis on LCAOMO Molecular Wave Functions" in J.Chem.Phys., 23, 1833 (1955).

Self Consistent perturbation theory is described in "Self Consistent Perturbation Theory" by G. Diercksen & R. McWeeny in J.Chem.Phys., 44, 3554 (1966)

The Rayleigh-Schrödinger Perturbation theory is developed in "Studies in Perturbation Theory I & II" in J.Molec.Spect., 13, 326 (1964) by P.-O. Löwdin.

11 OPEN SHELL SYSTEMS

11.1 UNPAIRED ELECTRONS

The methods and results we have discussed so far for the LCAOMO method all depend on the applicability of the equations derived in Chapters 5 and 9. The equations of Chapter 5 are limited to "closed shell" systems: the electronic structure is described by a single determinant of *doubly occupied* spatial orbitals. The vast bulk of organic and typical element inorganic molecules fall into this category but transition metal compounds and many transient species have unpaired electrons or "open shell" electronic structure. In order to include these important molecular species in our computational scheme we now turn to a sketch of the derivation of the relevant LCAOMO open shell RHF equations.

There are two possible approachs to unpaired electron systems within the LCAOMO framework, depending on the number of constraints to be imposed on the single determinant wave function. If we only ask that the wave function be a variational solution of the Schrödinger equation and satisfy the Pauli principle (the conditions of section 2.1) we obtain the "Different Orbitals for Different Spins" method. If, on the other hand, we insist that the approximate wave function represent a pure spin state the Open Shell RHF equation is obtained. Each of these methods is outlined below.

11.2 DIFFERENT ORBITALS FOR DIFFERENT SPINS* (DODS)

The wave function is written as a single determinant of n_α orbitals, ψ_i^α, which are occupied by electrons of α spin and n_β orbitals, ψ_j^β, in which electrons of spin β are placed

$$\Psi = \det\{\psi_1^\alpha \bar{\psi}_1^\beta \psi_2^\alpha \bar{\psi}_2^\beta \ldots \psi_{n_\alpha}^\alpha \psi_{n_\beta}^\beta\} \qquad (11.2.1)$$

where $n_\alpha \neq n_\beta$. The steps in deriving the variational equations for the ψ_i^α and ψ_j^β are completely analogous to those given in Chapter 4 for the closed shell case. The *spin orbitals* in (11.2.1) are all orthogonal since, as we shall see, the spatial orbitals ψ_i^α are all eigenfunctions of the same effective Hamiltonian and the same is true for the ψ_j^β. The spatial orbitals ψ_i^α and ψ_j^β do not have to be orthogonal since the corresponding spin orbitals are always orthogonal by spin integration. If we assume an LCAO expansion for both ψ_i^α and ψ_j^β the derivation of the equations satisfied by the linear coefficients follows the scheme:

i) Evaluate the total electronic energy associated with the function (11.2.1) using Slater's Rules
ii) Apply the variation principle to optimise the LCAO coefficients of each molecular orbital
iii) Hence obtain the equations satisfied by the LCAOMO expansions.

The energy expression obtained by use of Slater's Rules differs from the closed shell case in that each spatial orbital is occupied by only one electron; thus the factors of 2 do not appear in the DODS expression. The fact that $n_\alpha \neq n_\beta$ destroys the symmetry of the expression for $\mathsf{G}(\mathsf{R})$ - equation (5.2.12) - the effective potential is different for electrons of opposite spin. In fact the energy expression can be arranged in a form which

---

* This method was originally known as the UHF (Unrestricted Hartree-Fock) method. The DODS notation is clearer since it is also possible to remove the constraints of spatial symmetry on the wave function - we shall not go into these deep waters.

contains two "G matrices"; one for electrons of α spin and the other for β spin electrons. We write, in an obvious extension of the LCAOMO notation,

$$\begin{aligned} \boldsymbol{\psi}^{\alpha} &= \boldsymbol{\varphi}\mathsf{T}^{\alpha} \\ \boldsymbol{\psi}^{\beta} &= \boldsymbol{\varphi}\mathsf{T}^{\beta} \end{aligned} \tag{11.2.2}$$

with corresponding charge and bond order matrices

$$\begin{aligned} \mathsf{R}^{\alpha} &= \mathsf{T}^{\alpha}\,\mathsf{T}^{\alpha\dagger} \\ \mathsf{R}^{\beta} &= \mathsf{T}^{\beta}\,\mathsf{T}^{\beta\dagger} \end{aligned} \tag{11.2.3}$$

The energy expression is then

$$\begin{aligned} E = {} & \operatorname{tr}\mathsf{R}^{\alpha}\mathsf{H} + \tfrac{1}{2}\operatorname{tr}\mathsf{G}^{\alpha}\mathsf{R}^{\alpha} \\ & + \operatorname{tr}\mathsf{R}^{\beta}\mathsf{H} + \tfrac{1}{2}\operatorname{tr}\mathsf{G}^{\beta}\mathsf{R}^{\beta} \end{aligned} \tag{11.2.4}$$

where H has its usual meaning and

$$\begin{aligned} G^{\alpha}_{ij} &= \sum_{r,s=1}^{m} \{(R^{\alpha}_{rs} + R^{\beta}_{rs})(ij,rs) - R^{\alpha}_{rs}(ir,js)\} \\ G^{\beta}_{ij} &= \sum_{r,s=1}^{m} \{(R^{\alpha}_{rs} + R^{\beta}_{rs})(ij,rs) - R^{\beta}_{rs}(ir,js)\} \end{aligned} \tag{11.2.5}$$

Minimising (11.2.4) with respect to each column of T^{α} and T^{β} gives an effective Hamiltonian for each set of coefficients

$$\begin{aligned} \mathsf{H}^{F\alpha}\,\mathsf{T}^{\alpha} &= \mathsf{T}^{\alpha}\,\boldsymbol{\epsilon}^{\alpha} \\ \mathsf{H}^{F\beta}\,\mathsf{T}^{\beta} &= \mathsf{T}^{\beta}\,\boldsymbol{\epsilon}^{\beta} \end{aligned} \tag{11.2.6}$$

These two equations are coupled since

$$H^{F\alpha} = H + G^{\alpha}$$
$$H^{F\beta} = H + G^{\beta} \tag{11.2.7}$$

and G^{α}, G^{β} *both* depend on R^{α} and R^{β}.

It is readily seen that, provided $n_{\alpha} \neq n_{\beta}$ then $T^{\alpha} \neq T^{\beta}$ and the orbitals are indeed different orbitals for different spins. This conclusion follows easily by inspection of (11.2.5); the electrons of opposite spins experience different mean electron repulsion through the second set of terms in G^{α} and G^{β}. In the case that $n_{\alpha}=n_{\beta}$ in (11.2.4) both equations (11.2.6) collapse into the closed-shell RHF equation.

The computer implementation of the two coupled equations (11.2.6) involves no problems different in principle from the closed shell case. It is only necessary to carry out the iterative calculation for both sets of electrons in parallel and require both T^{α} and T^{β} to be self consistent. The use of non-orthogonal AO's replaces equations (11.2.6) by

$$H^{F\alpha} T^{\alpha} = S T^{\alpha} \epsilon^{\alpha}$$
$$H^{F\beta} T^{\beta} = S T^{\beta} \epsilon^{\beta} \tag{11.2.8}$$

and the methods of Chapter 9 are still applicable to the solution of these equations.

The description of molecular electronic structure given by a single configuration of these DODS orbitals has important applications in the interpretation and computation of electron spin properties which we discuss in section 11.4. However, the use of different orbitals for different spins in describing magnetic systems give a conceptual picture rather different from the usual one of a set of doubly occupied spatial orbitals plus an "outer" unpaired electron (or electrons). This latter picture has its theoretical justification in the Open Shell LCAOMO method which we now outline.

11.3 THE OPEN SHELL LCAOMO METHOD

The object of the open shell method is to replace equations (11.2.6) by a single equation. That is, to define a *single set* of orbitals for the whole electronic system which are not associated with any particular electron spin: to define a *single effective field* for each electron in place of the two fields defined by (11.2.7). We therefore use the open shell MO wave function

$$\Psi = \det\{\psi_1\bar{\psi}_1\psi_2\bar{\psi}_2 \cdots \psi_{n_c}\bar{\psi}_{n_c}\psi_{n_c+1}\} \qquad (11.3.1)$$

where, for definateness, we have assumed a single unpaired electron (in orbital ψ_{n_c+1}, α spin) "outside" a closed shell of n_c doubly occupied orbitals. We must now attempt to find an effective Hamiltonian which has solutions $\psi_1,\ldots\psi_{n_c}$ *and* $\psi_{n_c+1},\ldots\psi_m$; any virtual orbitals are pure bonus.

We now use the familiar LCAOMO expansion for all the orbitals

$$\boldsymbol{\psi} = \boldsymbol{\varphi}\mathsf{T} = (\mathsf{T}_c, \mathsf{T}_o) \qquad (11.3.2)$$

where the matrix T, of molecular orbital coefficients, has been partitioned into the closed shell columns T_c and the open shell columns T_o. Corresponding density matrices for the two types of orbital are

$$\begin{aligned} \mathsf{R}_c &= \mathsf{T}_c\mathsf{T}_c^\dagger \\ \mathsf{R}_o &= \mathsf{T}_o\mathsf{T}_o^\dagger \end{aligned} \qquad (11.3.3)$$

and equation (11.2.4) still holds for the total energy of the system (with an appropriate change of notation)

$$\begin{aligned} E &= 2\mathrm{tr}\,\mathsf{R}_c\mathsf{H} + \mathrm{tr}\,\mathsf{G}_c\mathsf{R}_c \\ &+ \mathrm{tr}\,\mathsf{R}_o\mathsf{H} + \tfrac{1}{2}\mathrm{tr}\,\mathsf{G}_o\mathsf{R}_o \end{aligned} \qquad (11.3.4)$$

The factor of 2 occurs because the closed shell orbitals are doubly occupied. In (11.3.4)

$$G_{c\,ij} = \sum_{r,s=1}^{m} (R_{c\,rs} + R_{o\,rs})[2(ij,rs)-(ir,js)]$$

and (11.3.5)

$$G_{o\,ij} = G_{c\,ij} - \tfrac{1}{2}\sum_{r,s=1}^{m} R_{o\,rs}(ir,js)$$

Application of the variation principle gives the equations satisfied by T_o and T_c:

$$\mathsf{H}^{Fc}\,\mathsf{T}_c = \mathsf{T}_c\,\boldsymbol{\epsilon}_c$$

$$\mathsf{H}^{Fo}\,\mathsf{T}_o = \mathsf{T}_o\,\boldsymbol{\epsilon}_o \qquad (11.3.6)$$

(assuming, for simplicity, orthogonal AO's). The derivation so far is precisely parallel to the DODS case. However, we have not yet used the fact that we wish to insist on the use of (11.3.2); the orbitals shall be eigenvectors of the same effective Hamiltonian, parts of the *same* T matrix, for which

$$\mathsf{H}^{F}\,\mathsf{T} = \mathsf{T}\,\boldsymbol{\epsilon} \qquad (11.3.7)$$

where H^F is yet to be found. The matrix H^F has to combine the properties of H^{Fc} *and* H^{Fo}. So, we must find a matrix which has the same eigenvectors, T_o, as H^{Fo} and gives zero when multiplying T_c and another matrix which has the eigenfunctions of H^{Fc} and gives zero when multiplying T_o. Any additive combination of two such matrices has the property (11.3.2) of H^F.

A brief digression to establish some of the properties of R_c and R_o sets up the appropriate machinery. The eigenvectors T of (11.3.7) are orthogonal and normalised,

$$\sum_{j=1}^{m} T_{ji}\,T_{jk} = \delta_{ik}$$

or,

$$\mathsf{T}^{\dagger}\,\mathsf{T} = \mathsf{1} \qquad (11.3.8)$$

and, since T_c and T_o are simply partitions of T, they have the same properties

$$T_c^\dagger T_c = 1 \qquad (11.3.9)$$

$$T_o^\dagger T_o = 1 \qquad (11.3.10)$$

Further, since T_o and T_c have no column of T in common, (11.3.8) ensures that

$$T_c^\dagger T_o = T_o^\dagger T_c = 0$$

These properties of orthogonality and normalisation of the T matrices define the action of the R matrices on the various T matrices:

$$R_c T_c = T_c T_c^\dagger T_c = T_c$$

$$R_o T_o = T_o T_o^\dagger T_o = T_o$$

$$R_c T_o = T_c T_c^\dagger T_o = 0$$

$$R_o T_c = T_o T_o^\dagger T_c = 0$$

Thus, the matrices $(1 - R_c)$, $(1 - R_o)$ have the convenient properties:

$$(1 - R_c)T_c = 0 \qquad (11.3.11)$$

$$(1 - R_c)T_o = T_o \qquad (11.3.12)$$

$$(1 - R_o)T_c = T_c \qquad (11.3.13)$$

$$(1 - R_o)T_o = 0 \qquad (11.3.14)$$

That is, they leave one set of orbital coefficients unchanged and *annihilate* the other set. We can now use relations (11.3.11) - (11.3.14) to form H^F from H^{Fo} and H^{Fc}.

Consider the matrix

$$(1 - R_o)H^{Fc}(1 - R_o) \tag{11.3.15}$$

Operating on T_c we have, using the above relations (11.3.11) - (11.3.14)

$$\begin{aligned}(1 - R_o)H^{Fc}(1 - R_o)T_c &= (1 - R_o)H^{Fc}(T_c - 0)\\ &= (1 - R_o)H^{Fc}T_c = (1 - R_o)T_c\,\epsilon_c\\ &= T_c\,\epsilon_c\end{aligned}$$

Operating on T_o the matrix (11.3.15) gives

$$(1 - R_o)H^{Fc}(1 - R_o)T_o = (1 - R_o)H^{Fc}(T_o - T_o) = 0$$

Thus, (11.3.15) has the same eigenvectors as H^{Fc} and gives zero when operating on the eigenfunctions of H^{Fo}. A similar analysis of the effect of

$$(1 - R_c)H^{Fo}(1 - R_c) \tag{11.3.16}$$

on T_c and T_o shows that

$$(1 - R_c)H^{Fo}(1 - R_c)T_o = T_o\,\epsilon_o$$

$$(1 - R_c)H^{Fo}(1 - R_c)T_c = 0$$

Therefore any matrix which is a linear combination of (11.3.15) and (11.3.16)

$$a(1 - R_o)H^{Fc}(1 - R_o) + b(1 - R_c)H^{Fo}(1 - R_c)$$

(a, b arbitrary constants) has both T_c and T_o as eigenvectors: precisely the requirement for H^F of (11.3.7). In fact it is possible to make an obvious extension of the above analysis to include the virtual (unoccupied) orbitals T_u (say) and to define a matrix H^F which has solutions T_c, T_o and T_u;

$$H^F = a(1 - R_o)H^{Fc}(1 - R_o) + b(1 - R_c)H^{Fo}(1 - R_c)$$
$$+ c(1 - R_u)(2H^{Fc} - H^{Fo})(1 - R_u) \qquad (11.3.17)$$

where

$$R_u = T_u T_u^\dagger$$

is a formal, virtual "density matrix".

Although the matrix defined by (11.3.17) has the same eigenvectors as H^{Fc} and H^{Fo}, it is clear that the arbitrariness in its definition - the numerical constants a, b, c - must lead to a certain arbitrariness in the eigenvalues: the eigenvalues of the open-shell RHF matrix are not unique. Since the eigenvectors are unique, an interpretation of the relation between the eigenvalues and the "molecular orbital energies" of photoelectron spectroscopy can be obtained by considering the "expectation values" of the eigenvectors with H^F: the quantities

$$T^{(i)\dagger} H^F T^{(i)}$$

where $T^{(i)}$ is a column of T. This has been done, with some rather surprising conclusions about the ordering of open and closed shell orbital energies, by McWeeny and Dodds.

The conversion of (11.3.17) into a form suitable for use with non-orthogonal orbitals gives equations which are cumbersome to write down but quite straightforward to program. The relations (11.3.11) - (11.3.14) hold when the orbital basis is orthogonal and, from an implementation point of view, the simplicity of the orthogonal form of (11.3.17) is sufficiently attractive to make it worth-while to store and use *both* orthogonal and non-orthogonal R matrices. The non-orthogonal R matrices are used (together with the molecular integrals over the non-orthogonal AO's) to form H^{Fc} and H^{Fo} which are then transformed to an orthogonal basis and the *orthogonal* R's used to form H^F. The computer implementation follows the following scheme, where we use the bar notation of Chapter 9 to denote matrices referring to an orthogonal basis:

i) Compute an orthogonalisation matrix X from S

ii) Make initial estimates of R_o, R_c, $\bar{R}_o$, $\bar{R}_c$

($\bar{R}_o = R_o$; $\bar{R}_c = R_c$ is not too bad for a starting point) - note that $R_u = 1 - R_o - R_c$.

iii) Using R_c and R_o, form G_c and G_o in one pass of the repulsion integral file (equation (11.3.5)), hence form H^{Fo} and H^{Fc}.

iv) Form $\bar{H}^{Fo} = X^{\dagger} H^{Fo} X$

and $\bar{H}^{Fc} = X^{\dagger} H^{Fc} X$

the orthogonal basis open and closed shell Hamiltonian matrices.

v) Using $\bar{R}_c$, $\bar{R}_o$, $\bar{R}_u$ and equation (11.3.17) form $\bar{H}^F$ from $\bar{H}^{Fo}$ and $\bar{H}^{Fc}$.

vi) Diagonalise $\bar{H}^F$, yielding $\bar{U}$ hence $\bar{T}$.

vii) From $\bar{T}_c$ and $\bar{T}_o$ form $\bar{R}_c$ and $\bar{R}_o$.

viii) Form the non-orthogonal eigenvector matrix

$T = X\bar{T}$

ix) Using T, form R_c and R_o

x) Check R_c and R_o for convergence,

finish or go to (iii) for further iterations.

This sketch of the implementation seems rather more involved than the closed shell case but all the additional steps are simple matrix multiplications and no new techniques are required. Small changes to the closed shell $G(R)$ routine enable the open and closed shell G matrices to be formed easily.

11.4 COMMENTS ON THE UNPAIRED ELECTRON METHODS

One of the most common applications of LCAOMO methods for unpaired electron systems is in the calculation and interpretation of EPR hyperfine coupling constants. The values of these quantities are determined largely by the "electron spin unbalance" at magnetic nuclei in the molecule: the excess density of electrons of one spin over electrons with the opposite spin at such a nucleus is the major factor involved. If the open shell method of 11.3 is used and it happens that the unpaired electron resides in an orbital with a node at the nucleus of interest (e.g. the planar CH_3 radical) then the computed hyperfine constant is, of course, zero. Using the DODS method the corresponding coupling constant is not computed to be zero because the unpaired electron has affected the "inner" orbitals by different amounts and a net "spin density" is predicted *even in the one-configuration approximation*. However, it is easy to show that the DODS wave function does not represent a pure spin state - it is not an eigenfunction of $\hat{S}^2$. It is therefore rather uncertain what value can be placed on computed spin properties using the DODS wave function. There are methods for generating a spin eigenfunction from a DODS single-determinant wave function but the resulting function loses the property of being a variational solution of the Schrödinger equation. Any attempt to use the DODS method together with the spin eigenfunction constraint takes us beyond the LCAOMO method into a MCSCF formalism.

It is usual to use the open shell method for the calculation of the electronic structure of, for example, transition metal complexes in which spin properties are not specifically sought.

11.5 APPLICATION TO BeH_2

There are no radicals or ions of the BeH_2 molecules of chemical interest so that our choice of species here is quite arbitrary. Nevertheless it is instructive to compare the α and β spin orbitals of the DODS calculation with each other and with the open shell orbitals for the same electronic system, BeH_2^+. In line with our interpretation of the spin eigenfunction condition as a *constraint* on the wave function, we see that the DODS wave function corresponds to a lower total energy for the system than the open shell wave function. The captions on the following Table describe the various matrices.

Table 11.1 Molecular Orbitals for BeH_2^+

DODS Wave Function

ε_i^α	-4.9223	-0.8889	-0.8650
	-0.0187	0.6740	0.6358
	-0.0187	0.6740	-0.6358
T^α	1.0027	-0.1950	0.0
	-0.0526	-0.2028	0.0
	0.0	0.0	0.0847
ε_i^β	-4.9283	-0.7194	
	-0.0205	0.1895	
	-0.0205	0.1895	
T^β	1.0032	-0.2924	
	-0.0520	0.5523	
	0.0	0.0	

Total Electronic Energy = -18.01183 a.u.

Open-Shell RHF Wave Function

ε_i	-4.9221	-0.7583	-0.6520
	-0.0067	0.4670	0.4658
	-0.0067	0.4670	-0.4658
T	0.9898	-0.1983	0.0
	0.0482	0.3837	0.0
	0.0	0.0	-0.4041

Total Electronic Energy = -18.00161 a.u.

SUGGESTIONS FOR FURTHER READING

"SCF Theory for Open Shells of Electronic Systems" by C.C.J. Roothaan in Rev.Mod.Phys., 32, 179 (1960) gives an alternative open-shell method in which two Hamiltonians are used. Roothaan's parameter a, b and c are related to the arbitrary constants a, b and c in (11.3.17) of course. Using $a = b = c = \frac{1}{2}$ in (11.3.17) gives $a = \frac{1}{2}$, $b = \frac{1}{2f}$, $c = \frac{1}{2(1-f)}$ for Roothaan's f, a, b, c. The derivation given here is based on the one in "A Self-Consistant Generalisation of Hückel Theory" by R. McWeeny in "Molecular Orbitals in Physics, Chemistry and Biology" (Academic Press 1964).

A discussion of Koopmans' Theorem for open-shell systems is given in "Orbital Energies and Koopmans' Theorem in Open-Shell Hartree-Fock Theory" by J.L. Dodds and R. McWeeny, Chem.Phys.Letters, 13, 9 (1972).

12 THE USE OF MOLECULAR SYMMETRY

12.1 MOLECULAR SYMMETRY AND AO'S

Many molecules, radicals and ions of chemical interest are symmetrical in the sense of containing a mirror plane or some other obvious element of spatial symmetry. The existence of planes and axes of symmetry in a molecule enable considerable savings to be made throughout the process of computing molecular wave functions in the orbital model. The two main areas of computational work - molecular integral evaluation and the iterative solution of the RHF equation - can both be speeded up considerably by elementary techniques based on the minimum of formal group representation theory. The main practical use of molecular symmetry can be easily developed from first principles

An operation which takes an object from a reference position into a position which is indistinguishable from that reference (with respect to the fixed surroundings) is called a *symmetry operation*. If one or more such operations are discovered for a molecule and new operations found by exhaustive re-application of these operations then the set of symmetry operations generated necessarily form a group: *a symmetry group of the molecule*. In the case of molecular systems the relevant symmetry group is the one whose operations permute like nuclei consistent with the maintenance of the overall geometry of the molecule. In more formal terms it is the group of symmetry operations on the coordinates of the nuclei *which leave the molecular Hamiltonian invariant*. This definition is not unique of course; any sub-group of the molecular symmetry group satisfies these conditions.

If the AO's with which we describe the electronic structure of the molecule reflect the symmetry of the nuclear arrangement - like atoms have like orbitals centred on them - then there are redundancies in an orbital basis calculation. Some of the MO coefficients (elements of $\mathbf{T}$) must be related and, more important, some molecular integrals must be related, by symmetry. In our BeH_2 example the electron repulsion integrals (11,11) and (22,22) must be the same and the squares of the coefficients of the two hydrogen orbitals, ϕ_1 and ϕ_2, in any MO must be identical. We can make use of this type of symmetry restriction in the MO calculation.

When a symmetry operation is applied to the AO's of a molecule the orbitals of one atom are sent into the orbitals of a like atom - either a simple permuation of the orbitals or a linear combination of the original AO's. Thus, if $\boldsymbol{\varphi}$ is the basis of AO's, we can express the effect of applying a symmetry operation $\hat{G}_i$ as

$$\hat{G}_i \boldsymbol{\varphi} = \boldsymbol{\varphi}^{(i)} \tag{12.1.1}$$

where $\boldsymbol{\varphi}^{(i)}$ is the transformed set. The transformation is obviously linear, so that equation (12.1.1) can be replaced by a matrix equation by collecting the coefficients of the transformation as columns of a matrix $\mathbf{D}^{(i)}$:

$$\boldsymbol{\varphi}^{(i)} = \boldsymbol{\varphi}\mathbf{D}^{(i)} \tag{12.1.2}$$

This simple equation is enough to enable us to examine the effect of molecular symmetry on the most time consuming step in the molecular calculation: the computation, storage and retrieval of the molecular integrals. The use of molecular symmetry to simplify the RHF equation needs more theoretical background than (12.1.2) and is taken up in a later section.

12.2 MOLECULAR SYMMETRY AND MOLECULAR INTEGRALS

In the interests of clarity, throughout this section we treat only those molecular symmetry groups whose symmetry elements *permute* the AO's of the molecule*; we specifically exclude those symmetry operations which induce linear transformations among the AO's. As soon as the method of using molecular symmetry is made clear this restriction will be dropped. Carrying the full treatment in the early stages of the development rather clouds the essential ideas with spurious generality.

Since the matrices $\mathbb{D}^{(i)}$ of (12.1.2) effect only a permutation among the AO's, each column of any $\mathbb{D}^{(i)}$ contains only one non-zero element (unity) and the information in the matrix can be summarised much more compactly as a list of numbers representing the permutation. The effect on the AO's of BeH_2 of reflection in the plane passing through the Be atom, perpendicular to the internuclear axis, is represented by

$$\boldsymbol{\varphi}^{(i)} = (\phi_1,\phi_2,\phi_3,\phi_4,\phi_5)\begin{pmatrix} 0 & 1 & 0 & 0 & 0 \\ 1 & 0 & 0 & 0 & 0 \\ 0 & 0 & 1 & 0 & 0 \\ 0 & 0 & 0 & 1 & 0 \\ 0 & 0 & 0 & 0 & -1 \end{pmatrix} \qquad (12.2.1)$$

where

$$\boldsymbol{\varphi}^{(i)} = (\phi_2,\phi_1,\phi_3,\phi_4,-\phi_5)$$

In an obvious notation (12.2.1) is easily summarised by:

$$\begin{matrix} 1 & & 2 \\ 2 & & 1 \\ 3 & \rightarrow & 3 \\ 4 & & 4 \\ 5 & & -5 \end{matrix} \qquad (12.2.2)$$

Working with a specific example will show the features of the

---

* Strictly, it is the *combination* of the particular symmetry group and the chosen set of AO's which has the permutation property.

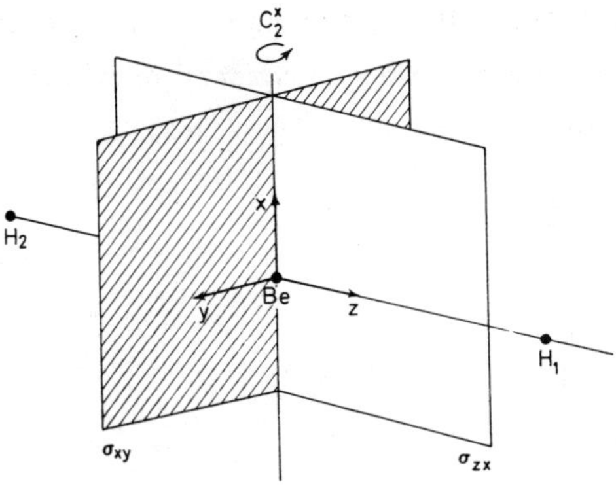

Figure 12.1 The symmetry elements used for BeH_2

method in the most direct way. The BeH_2 molecule with the AO's used earlier was chosen with this application in mind since this example shows all the features of the general case. The group of symmetry operations comprising:

i) The identity operation (E);

ii) Rotation by π about the x axis (C_2^x);

iii) Reflection in the xy plane (σ^{xy});

iv) Reflection in the zx plane (σ^{zx});

(referring to co-ordinate system of Fig. 12.1) fulfils the conditions of leaving the Hamiltonian invariant *and* merely permutes the five AO's. Using the permutation notation of (12.2.2), the effect of all these symmetry operations can be summarised as follows:

E	C_2^x	σ^{xy}	σ^{zx}
1	2	2	1
2	1	1	2
3	3	3	3
4	4	4	4
5	-5	-5	5

Each column gives the permutation induced among the five AO's by the operation at the head of the column. The 2p AO changes sign under the action of some of the symmetry operations and this has been indicated in the obvious way in row 5 of the table.

We now regard this table of permutations as a matrix M whose elements are defined by:

M_{ij} is the number of the orbital into which orbital ϕ_i is sent by the j'th symmetry operation

The numbering system refers of course to the original ϕ_1 - ϕ_5 for the AO's and 1 through 4 for E, C_2^x, σ^{xy}, σ^{zx}. In the restricted case of orbital permutations only, the matrix M clearly contains all the symmetry information about the AO's. The use of this scheme is quite general and can be used for any molecular symmetry and any AO set. This information, in this form, will enable us to make considerable savings during molecular integral computations.

We begin by considering the one electron integrals

$$H_{ij} = \int dr\phi_i(r)\hat{h}\phi_j(r).$$

The Hermitian symmetry of $\hat{h}$ ensures that $H_{ij} = H_{ji}$ independent of any molecular symmetry and so we ignore this trivial symmetry. It is intuitively obvious from Fig. 12.1 that $H_{11} = H_{22}$ for BeH_2 since ϕ_1 and ϕ_2 are identical orbitals in identical environments. Similar reasoning shows that H_{13} and H_{23} must be identical. These equalities, and others more difficulty to spot by inspection, can easily be obtained by systematic examination of the matrix M (and of course, the knowledge that $\hat{h}$ is invariant with respect to any symmetry operation).

Examination of the first *row* of M, which summarises the fate of orbital ϕ_1 under the action of the four symmetry operations, shows that ϕ_1 is always sent into itself or ϕ_2: row 1 of M contains only 1's and 2's. Therefore any integrals involving only ϕ_1 or only ϕ_2 must be identical since these orbitals are *equivalent* in the molecular symmetry considered. Similar examination of row 3 of M shows that ϕ_3 is left unchanged by all the operations. Thus any integral involving ϕ_1 and ϕ_3 must be identical to an integral of the same form involving ϕ_2 and ϕ_3. Working in this way with the first four rows of M and collecting results, we see that, of the ten potentially different integrals H_{ij} involving the first four orbitals, only seven are actually distinct due to molecular symmetry. These can be chosen as H_{11}, H_{12}, H_{13}, H_{33}, H_{14}, H_{34} and H_{44}, leaving H_{22}, H_{23} and H_{24} to be determined by symmetry. This is convenient but not spectacular; a saving of three integrals.

Examination of the fifth row of M - the effect of the symmetry operations on the 2p AO - reveals a new phenomenon. If we ask for the one electron integrals equivalent to H_{15} we find that the operations which send ϕ_1 into ϕ_2 sends ϕ_5 into $-\phi_5$. Checking the physical interpretation of this fact with Fig. 12.1 shows the meaning: integral H_{25} has the same magnitude as H_{15} but *opposite sign*. The negative lobe of the 2p orbital overlaps with the 1s hydrogen AO in one case and the positive lobe in the other. We need, therefore, to save the *sign* of the product of the elements of $\mathbf{M}$ as well as the orbital numbers. In the second column of $\mathbf{M}$ we see that the operation C_2^x, which sends ϕ_4 into itself, sends ϕ_5 to $-\phi_5$. This suggests that $H_{45} = -H_{45}$ i.e. $H_{45} = 0$ and the physical interpretation confirms that this integral is zero; positive and negative regions of overlap cancelling. Similar reasoning shows $H_{35} = 0$. Summarising the results, the integrals:

$$H_{11},\ H_{12},\ H_{13},\ H_{33},\ H_{14},\ H_{34},\ H_{44},\ H_{15},\ H_{55}$$

must be computed, and the rest:

$$H_{22},\ H_{23},\ H_{24},\ H_{25},\ H_{35},\ H_{45}$$

are either zero or can be obtained from the computed results. This gives a saving of about one third. The small amount of time taken in scanning $\mathbf{M}$ must be set against this saving.

In order to develop an effective algorithm for automating this procedure, we need to take decisions on one or two points which present no difficulty to the reader but are not so obvious to a computer.

i) In applying the permutations of $\mathbf{M}$ to the indices i and j of a particular integral H_{ij}, it is possible that *both* $H_{i'j'}$ and $H_{j'i'}$ are formed ($\phi_{i'}$, $\phi_{j'}$ are the orbitals equivalent to ϕ_i, ϕ_j). The latter integrals are trivially equal, so we must ensure that i' and j' are *ordered* before any comparisons are performed. Using $i' \geqslant j'$ falls in line with our earlier conventions.

ii) In most cases an integral H_{ij} will be transformed into $H_{i'j'}$ *several times* by the permutations of $\mathbf{M}$ (c.f. H_{12} in BeH_2). Also, if H_{ij} is transformed into $H_{i'j'}$, then, when the time comes to process $H_{i'j'}$, H_{ij} will be generated again. It is necessary to incorporate a method for *discarding repetitions*. The second type of repetition can be avoided by placing the H_{ij} in some standard order - say the order of increasing [ij] - and referring to *later* or *earlier* occurrences of H_{ij}

In short, we work with an ordering convention $i \geqslant j$ for the labels and order of increasing [ij] for the values: precisely analogous to the two electron integral conventions of Chapter 8. Both of these conventions are generated by

```
      DO   1  I=1,M
      DO   1  J=1,I
      .
      .
      H(I,J) = . . . .
    1 CONTINUE
```

Thus $I \geqslant J$ and the integrals H(I,J) occur in order of increasing [IJ].

With these conventions we can now list an algorithm for the use of M to obtain maximum saving of resources due to symmetry during the computation of one electron integrals:

1. Choose the first pair of labels (i,j) = (1,1).
2. Apply the permutations along the i'th and j'th rows of M to i and j respectively. Collect the set of pairs of labels {(i',j')} produced and save the sign of each product i' j'.
3. Put the labels i', j' in each member of the set {(i',j')} in the order $i' \geqslant j'$.
4. If any of the pairs (i',j') occurs *earlier* than (i,j) - in the sense of (ii) above - discard the whole set and go to 9 (By the group property the whole set must have been generated earlier). If not, continue.
5. If any of the set {(i',j')} is the same as (i,j) but of opposite sign, discard the whole set since all are zero then go to 9. If not, continue.
6. Discard any repetitions among (i,j) and {(i',j')}.
7. Compute and save H_{ij}.
8. Set the $H_{i'j'}$ for the remaining {(i',j')} equal to H_{ij} (with sign from step 2) and store them.

9. Go on to the next pair (i,j).
10. If j is less than the number of orbitals go to 2, otherwise finish.

An annotated program fragment is given below to do the same task. A FUNCTION called OEI has been assumed to exist which does the purely numerical work of integral evaluation - an implementation of the formulae in Appendix A.

This algorithm, and the computer implementation of it, are in no way restricted to the example we have used to develop it or to the restricted set of AO's used. The only limitation is that the symmetry operations must be chosen to avoid linear combinations among the AO's. The more symmetrical the molecule the greater the saving, the permanganate ion showing the method to devastating effect: a basis of 39 AO's would require the computation of 780 one electron integrals, the use of symmetry reduces this number to 76.

Having established the symmetry procedure for one electron integrals* - two label quantities - it is straightforward to go on and examine the use of symmetry in the computation of the electron repulsion integrals - four label quantities. The permutations defined gy the rows of $\mathbb{M}$ are this time applied to the *four* labels of an electron repulsion integral and similar ordering conventions enable an analogous algorithm to be developed. The ordering conventions of Chapter 8,

$$i \geq j; \quad k \geq \ell; \quad [ij] \geq [k\ell]$$

for the labels, and the values are placed in order of increasing [[ij][kℓ]]. The one electron algorithm listed earlier can now be applied simply by changing references to pairs (i,j) to sets of four (i,j,k,ℓ). The two electron algorithm will not be set out in detail because it is so similar to the earlier, one

---

* Exactly the same algorithm can be used to evaluate the overlap integrals or, better, H_{ij} and S_{ij} can both be computed at the same time.

```
      INTEGER    SIGN
      DIMENSION MTRAN(10,24),ISAVE(24),JSAVE(24)
      DIMENSION ISIGN(24),ISP(24),H(10,10)
      IREAD=2
      READ(IREAD,100) N,MMAX
  100 FORMAT (24I3)
      DO 99 I=1,N
   99 READ(IREAD,100)  (MTRAN(I,J),J=1,MMAX)
C   LOOPS OVER I AND J
      DO 1 I=1,N
      DO 1 J=1,I
      IJ=I*(I-1)/2+J
      IPR=I*J
C   PUT INFORMATION ABOUT CURRENT H(I,J)
C   AT HEAD OF LIST OF EQUIVALENT INTEGRALS
      KOUNT=1
      ISAVE(1)=I
      JSAVE(1)=J
      ISIGN(1)=1
      ISP(1)=IJ
C   LOOP OVER GROUP OPERATIONS
      DO 2 M=1,MMAX
      IT=MTRAN(I,M)
      JT=MTRAN(J,M)
      ITPR=IT*JT
      SIGN=1
      IF(ITPR.LT.0)  SIGN=-1
C   ORDER TRANSFORMED LABELS AND SAVE SIGN
      IT=IABS(IT)
      JT=IABS(JT)
      IF(IT.GE.JT)   GO TO 3
      ID=IT
      IT=JT
      JT=ID
    3 IJT=IT*(IT-1)/2+JT
C   CHECK IF THIS H(I,J) HAS BEEN COMPUTED
C     EARLIER
      IF(IJ-IJT) 4,5,1
C   CHECK IF H(I,J) IS ZERO
    5 IF(IPR+ITPR)  2,1,2
C   DISCARD REPITITIONS
    4 DO 6 KC=1,KOUNT
      IF(ISP(KC) .EQ.IJT) GO TO 2
    6 CONTINUE
      KOUNT=KOUNT+1
      ISP(KOUNT) =IJT
      ISAVE(KOUNT)=IT
      JSAVE(KOUNT)=JT
      ISIGN(KOUNT)=SIGN
    2 CONTINUE
C   COMPUTE ONE OF THE INTEGRAL VALUES
      AA=OEI(I,J)
C   SET THE REST EQUAL (WITH SIGN)
      DO 7 KC=1,KOUNT
      IT=ISAVE(KC)
      JT=JSAVE(KC)
      H(IT,JT)=ISIGN(KC)*AA
    7 H(JT,IT)=H(IT,JT)
    1 CONTINUE
      STOP
      END
```

```
      INTEGER SIGN
      DIMENSION MTRAN(10,24),ISAVE(24),JSAVE(24)
      DIMENSION KSAVE(24),LSAVE(24),ISIGN(24),AISP(24)
      DIMENSION II(200),JJ(200),KK(200),LL(200),VALUE(200)
      DATA ZERO,NN/0.0,200/
C   STATEMENT FUNCTIONS FOR  (IJ) AND ((IJ)(KL))
      IJ(I,J)=I*(I-1)/2+J
      P4(I,J,K,L)=0.5*FLOAT(IJ(I,J))* FLOAT(IJ(I,J)-1)
     *    +FLOAT(IJ(K,L))
      IREAD=2
      READ(IREAD,100) N,MMAX
  100 FORMAT(24I3)
C   MTRAN IS 'M' OF TEXT
      DO 99 I=1,N
   99 READ(IREAD,100) (MTRAN(I,J),J=1,MMAX)
      NFILE=10
      IEND=0
      IM=0
C   LOOPS OVER 'DISTINCT' INTEGRAL LABELS
      DO 1 I=1,N
      DO 1 J=1,I
      DO 1 K=1,I
      LTOP=K
      IF(I.EQ.K)  LTOP=J
      DO 1 L=1,LTOP
      IF(L.EQ.N) IEND=1
      P=P4(I,J,K,L)
      KOUNT=1
C   PUT (IJ,KL) INFORMATION AT HEAD OF LIST
C   OF POINT GROUP EQUIVALENT INTEGRALS
      ISAVE(1)=I
      JSAVE(1)=J
      KSAVE(1)=K
      LSAVE(1)=L
      AISP(1)=P
      ISIGN(1)=1
      PROD=FLOAT(I*J)*FLOAT(K*L)
C   LOOP OVER GROUP OPERATIONS
      DO 2 M=1,MMAX
      IT=MTRAN(I,M)
      JT=MTRAN(J,M)
      KT=MTRAN(K,M)
      LT=MTRAN(L,M)
      TPROD=FLOAT(IT*JT)*FLOAT(KT*LT)
      SIGN=1
       IF(TPROD.LT.ZERO) SIGN=-1
C   ORDER TRANSFORMED LABELS
      CALL ORDER(IT,JT,KT,LT)
      PT=P4(IT,JT,KT,LT)
C   CHECK IF CURRENT INTEGRALS HAVE BEEN
C    COMPUTED EARLIER
      IF(P-PT) 3,4,1
C   CHECK IF (IJ,KL) IS ZERO
    4 IF(PROD+TPROD) 2,1,2
C   DISCARD REPITITIONS
    3 DO 5 KC=1,KOUNT
      IF(AISP(KC).EQ.PT)  GO TO 2
    5 CONTINUE
      KOUNT=KOUNT+1
C   SAVE EQUIVALENT LABELS AND SIGN
      AISP(KOUNT)=PT
      ISAVE(KOUNT)=IT
      JSAVE(KOUNT)=JT
      KSAVE(KOUNT)=KT
      LSAVE(KOUNT)=LT
      ISIGN(KOUNT)=SIGN
    2 CONTINUE
C   COMPUTE ONE OF EQUIVALENT INTEGRALS
      AA=ERI(I,J,K,L)
```

```
C   SET THE REST EQUAL (WITH SIGN)
      DO 6 KC=1,KOUNT
      IM=IM+1
      II(IM)=ISAVE(KC)
      JJ(IM)=JSAVE(KC)
      KK(IM)=KSAVE(KC)
      LL(IM)=LSAVE(KC)
      VALUE(IM)=ISIGN(KC)*AA
      IF(IEND.NE.0)   GO TO 8
      IF(IM.LT.NN)   GO TO 6
C   WRITE OUT THE BUFFER WHEN IT'S FULL
    8 WRITE(NFILE) IM,IEND,II,JJ,KK,LL,VALUE
      IM=0
    6 CONTINUE
    1 CONTINUE
      STOP
      END
```

```
      SUBROUTINE ORDER (I,J,K,L)
      I=IABS(I)
      J=IABS(J)
      K=IABS(K)
      L=IABS(L)
      IF(I-J) 1,2,2
    1 M=I
      I=J
      J=M
    2 IF(K-L) 3,4,4
    3 M=K
      K=L
      L=M
    4 IF(I-K) 6,5,7
    5 IF(J-L) 6,7,7
    6 M=I
      I=K
      K=M
      M=J
      J=L
      L=M
    7 RETURN
      END
```

electron, case. It is, however, worth listing a program fragment to perform the task since the method of storing electron repulsion integrals is different from the simple matrix method used for the elements of H. Also the "integer packing" process is used in the sample program in Appendix C which we have not yet met. Again, it is assumed that FUNCTION ERI is available to perform the numerical work of integral evaluation. The savings made by using the information in M during the computation of the two electron integrals are much more important than those involved in the computation of H because of the size of $m\mathcal{S}$ compared to $\frac{1}{2}m(m+1)$. To quote the AO basis permanganate calculation again, the number of electron repulsion integrals to be computed is cut from 304,590 to 15,132. To illustrate the form of the output two electron integral file for BeH_2 the electron repulsion integrals - as produced by a program similar to our example - are given below. For clarity each *computed* integral is underlined and the equivalent integrals follow. The destruction of the "natural" order of the integrals in the file is apparent from the listing.

This method ensures that only those molecular integrals which are *essentially different with respect to the chosen molecular symmetry group* are computed and so ensures maximum saving of resources for that particular symmetry group.

There are two obvious ways in which further savings of computer facilities can be made over and above the ones outlined.

i) The time involved in processing a field of electron repulsion integrals held on an external medium would be greatly reduced by *storing* only the essentially different integrals. The required storage space would also be drastically cut. If we can develop an algorithm for processing a file containing only the symmetry distinct integrals - analogous to the treatment of the trivial identities discussed in Chapter 8 - we can *compute and store* only these integrals.

1	1	1	1	0.62203	2	2	2	2	0.62203	2	1	1	1	0.03056
2	2	2	1	0.03056	2	1	2	1	0.00308	2	2	1	1	0.19969
3	1	1	1	0.03619	3	2	2	2	0.03619	3	1	2	1	0.00421
3	2	2	1	0.00421	3	1	2	2	0.03175	3	2	1	1	0.03175
3	1	3	1	0.00981	3	2	3	2	0.00981	3	2	3	1	0.00935
3	3	1	1	0.38938	3	3	2	2	0.38938	3	3	2	1	0.05008
3	3	3	1	0.13782	3	3	3	2	0.13782	3	3	3	3	2.29207
4	1	1	1	0.27833	4	2	2	2	0.27833	4	1	2	1	0.01824
4	2	2	1	0.01824	4	1	2	2	0.13394	4	2	1	1	0.13394
4	1	3	1	0.02345	4	2	3	2	0.02345	4	1	3	2	0.02141
4	2	3	1	0.02141	4	1	3	3	0.26168	4	2	3	3	0.26168
4	1	4	1	0.14141	4	2	4	2	0.14141	4	2	4	1	0.08824
4	3	1	1	0.08479	4	3	2	2	0.08479	4	3	2	1	0.01061
4	3	3	1	0.02445	4	3	3	2	0.02445	4	3	3	3	0.35463
4	3	4	1	0.05627	4	3	4	2	0.05627	4	3	4	3	0.06248
4	4	1	1	0.32421	4	4	2	2	0.32421	4	4	2	1	0.03223
4	4	3	1	0.04162	4	4	3	2	0.04162	4	4	3	3	0.49059
4	4	4	1	0.19452	4	4	4	2	0.19452	4	4	4	3	0.10478
4	4	4	4	0.34728	5	1	1	1	-0.34225	5	2	2	2	0.34225
5	1	2	1	-0.01808	5	2	2	1	0.01808	5	1	2	2	-0.11022
5	2	1	1	0.11022	5	1	3	1	-0.02244	5	2	3	2	0.02244
5	1	3	2	-0.01893	5	2	3	1	0.01893	5	1	3	3	-0.23680
5	2	3	3	0.23680	5	1	4	1	-0.16120	5	2	4	2	0.16120
5	1	4	2	-0.07625	5	2	4	1	0.07625	5	1	4	3	-0.05162
5	2	4	3	0.05162	5	1	4	4	-0.19194	5	2	4	4	0.19194
5	1	5	1	0.19601	5	2	5	2	0.19601	5	2	5	1	-0.05952
5	3	1	1	-0.00933	5	3	2	2	0.00933	5	3	3	1	-0.00108
5	3	3	2	0.00108	5	3	4	1	-0.00436	5	3	4	2	0.00436
5	3	5	1	0.00751	5	3	5	2	0.00751	5	3	5	3	0.00527
5	4	1	1	-0.10789	5	4	2	2	0.10789	5	4	3	1	-0.00229
5	4	3	2	0.00229	5	4	4	1	-0.04296	5	4	4	2	0.04296
5	4	5	1	0.06864	5	4	5	2	0.06864	5	4	5	3	0.01005
5	4	5	4	0.07581	5	5	1	1	0.36767	5	5	2	2	0.36767
5	5	2	1	0.03285	5	5	3	1	0.04055	5	5	3	2	0.04055
5	5	3	3	0.46759	5	5	4	1	0.20583	5	5	4	2	0.20583
5	5	4	3	0.10131	5	5	4	4	0.34684	5	5	5	1	-0.21398
5	5	5	2	0.21398	5	5	5	5	0.37427					

ii) We must drop the restriction made at the beginning of this section and include symmetry operations which induce linear combinations among the AO's. Molecules with fivefold or sixfold axes (nickelocene, benzene) can then be given adequate treatment.

It is obvious that one solution to the problem posed by (i) would be to store the distinct integrals and simply apply the algorithm given above "in reverse" when processing the integrals. That is, re-generate all the integrals equivalent to the stored one each time the integrals are used:

1. Take the first value of H_{ij}, labels (i,j)
2. Apply the permutations of M to generate the equivalent set {(i',j')} - with sign
3. Order {(i',j')} and discard repetitions
4. Use H_{ij} and the $H_{i'j'}$
5. Take the next distinct integral H_{ij} and go to 2.

In practice although the amount of time spent scanning M is small compared to the computation time it is not negligible, and experience has shown that the procedure sketched above is actually *considerably slower* than reading and processing the full file of integrals (in the important two electron integral case).

For those applications in which each electron repulsion integral appears essentially independently - the VB/CI method - there is, at present, no convenient method of storing only the distinct integrals which effects a net saving of resources. In the LCAOMO method the situation is quite different; a simple and effective means of using a file of symmetry distinct electron repulsion integrals (and one electron integrals) has been developed and is in use in some computing laboratories. The essential point on which this method hinges is the fact that the electron repulsion integrals only occur in a *particular combination* in $\mathsf{G}(\mathsf{R})$. The matrix $\mathsf{G}(\mathsf{R})$ has the *symmetry of a one electron operator* and so the one electron algorithm can be applied to the two electron integrals *in the MO calculation*.

12.3 STORAGE COMPRESSION USING MOLECULAR SYMMETRY

The development of a method of storing only the symmetry distinct molecular integrals is best illustrated through an example - BeH_2 again! Consider the matrix which contains only the distinct one electron integrals and zeroes elsewhere:

$$\begin{pmatrix} H_{11} & H_{12} & H_{13} & H_{14} & H_{15} \\ H_{12} & 0 & 0 & 0 & 0 \\ H_{13} & 0 & H_{33} & H_{34} & 0 \\ H_{14} & 0 & H_{34} & H_{44} & 0 \\ H_{15} & 0 & 0 & 0 & H_{55} \end{pmatrix}$$

If we take each non-zero element H_{ij} in turn, apply the permutations of M to (i,j), do *not* discard repetitions and add the results we obtain the following matrix:

$$\begin{pmatrix} 2H_{11} & 4H_{12} & 2H_{13} & 2H_{14} & 2H_{15} \\ 4H_{12} & 2H_{22} & 2H_{13} & 2H_{14} & -2H_{15} \\ 2H_{13} & 2H_{13} & 4H_{33} & 4H_{34} & 0 \\ 2H_{14} & 2H_{14} & 4H_{34} & 4H_{44} & 0 \\ 2H_{15} & -2H_{15} & 0 & 0 & 4H_{55} \end{pmatrix}$$

This matrix does have the correct molecular symmetry properties but some of the elements have the wrong values - due to the fact that not all the permutations of M generate different $H_{i'j'}$. However, each set of symmetry equivalent H_{ij}'s are multiplied by the same factor: *the number of times element H_{ij} is sent into itself by the permutations of* M. So, if we compute and store the symmetry distinct integrals *divided by the number of times they are sent into themselves by the operations of the group* then this simple procedure generates the correct matrix.

To apply this procedure to the G matrix it is necessary to recall the definition of a typical element G_{ij} in terms of the repulsion integrals and the elements of R.

$$G_{ij} = \sum_{r,s} R_{rs}[2(ij,rs) - (ir,js)] \tag{12.3.1}$$

An examination of the effect of the permutations of M on the elements of R and on the repulsion integrals gives the effect of these transformations on the G matrix. For simplicity we restrict our attention to the first term in G_{ij},

$$2R_{rs}(ij,rs) \tag{12.3.2}$$

and assume that a typical permutation is given by;

$$\phi_i' = \sigma_{ii'}\phi_i$$
$$\phi_j' = \sigma_{jj'}\phi_j$$
$$\phi_k' = \sigma_{kk'}\phi_k$$
$$\phi_\ell' = \sigma_{\ell\ell'}\phi_\ell$$

where the $\sigma_{ii'}$ etc. are ±1 - the sign of the relevant element of M. There will be an equivalent contribution to $G_{i'j'}$ of the same form as (12.3.2) since:

$$R_{r's'} = \sigma_{rr'}\sigma_{ss'}R_{rs}$$

and

$$(i'j',r's') = \sigma_{ii'}\sigma_{jj'}\sigma_{rr'}\sigma_{ss'}(ij,rs)$$

giving

$$2\sigma_{rr'}\sigma_{ss'}R_{rs}\sigma_{ii'}\sigma_{jj'}\sigma_{rr'}\sigma_{ss'}(ij,rs)$$

or, since $\sigma_{rr'}^2 = \sigma_{ss'}^2 = 1$,

$$\sigma_{ii'}\sigma_{jj'}[2R_{rs}(ij,rs)] \tag{12.3.3}$$

precisely the effect of applying the permutations of M to the partial G matrix obtained from *only the distinct electron repulsion integrals*. Just as in the discussion of H, if we want to avoid the bother of discarding repetitions, it is necessary to divide each repulsion integral by the number of times it is sent into itself by the permutations of M.

In practice, we form the matrix

$$\mathsf{H}^F = \mathsf{H} + \mathsf{G(R)}$$

from the scaled distinct elements of H and the scaled distinct repulsion integrals and then apply the permutations of M to the whole matrix and sum the results. The program to perform this last operation is simply tacked on to the end of the $\mathsf{G(R)}$ routine given in Chapter 8. Of course, the "trivial" symmetries - the allowed permutations among i, j, k and ℓ of one $(ij,k\ell)$ - must still be taken care of by the original method. The fact that no changes have to be made in the routine for $\mathsf{G(R)}$ formation illustrates the flexibility of our two electron integral storage method: the program simply takes those integrals in the file and forms the full or partial $\mathsf{G(R)}$. There is no need to inform the program that we are using a shorter file or that the integrals have been scaled. The matrix M, used to generate the full H^F matrix, must be the same as the one used in the computation of H and the $(ij,k\ell)$. The program GOFR in Appendix C illustrates the computer implementation of these ideas and the routine DGLST illustrates the calculation and storage of the distinct, scaled, repulsion integrals.

12.4 EXTENSION TO GENERAL LINEAR TRANSFORMATIONS

There is no difficulty of *principle* involved in extending the treatment of "permutation only" operations, given in section 12.2, to symmetry operations which induce linear combinations of the AO's used to describe a molecule. The problem is simply one of book-keeping: the necessity of storing all the relevant linear combination coefficients and the many small groups of molecular integrals among the symmetry related AO's. A symmetry operation which causes linear combinations among n orbitals on each of two atoms necessitates the use of $(2n)^4$ electron repulsion integrals in a temporary store. The complications involved in designing a *general system* for any molecular symmetry with any number and type of AO's are enormous. For applications outside the MO framework (which again is a special case) it has been found convenient to use a modified form of the method of 12.2 in which the "permutation symmetries" are used and the "linear combination symmetries" are not. In this method we simply put a marker in $\mathbb{M}$ (zero will serve admirably since it is not the number of an orbital) if the orbital is sent into a linear combination of the AO's by that particular symmetry operation. A trivial logical addition to our program implementation can be made which traps these markers and *always computes* an integral whose labels contain at least one such marker. This partial use of symmetry does not, for example, make use of the symmetry relation between the two in-plane 2p orbitals on each carbon atom in benzene or similar relations among the angle dependent AO's when the molecular symmetry axis is not two- or four-fold. It might be objected that the in-plane orbitals of benzene could be *chosen* so that they *are* permuted by the symmetry operations - this device is used by many authors in obtaining qualitative results from symmetry considerations - "radial" and "tangential" 2p orbitals in benzene would be a suitable choice. This merely begs the question from a computational point of view because the molecular integral formulae (for GTF's) depend for their simplicity on the use of an overall global co-ordinate system and therefore *parallel local atomic axis systems*. The computation of the linear transformation effects would merely be done *in another place* in the program.

The MO method again has the advantage of having all the two electron integrals conveniently collected in the one electron package $\mathbb{G}(\mathbb{R})$. This matrix has the transformation properties of a one electron operator and it is quite feasible to save the associated linear transformation coefficients since there are relatively few of them. The matrices $\mathbb{D}^{(i)}$ of (12.1.2) can always be chosen to be unitary: the permutation matrices similar to (12.2.1) are the simplest type of unitary transformation. The derivation of (12.3.3) from (12.3.2) is, in effect, the use of a set of unitary 1×1 matrices, +1 and -1, (the $\sigma_{ii'}$) in a theorem which can be derived for unitary m×m matrices. If we change the quantities in the derivation to bold-face and replace the $\sigma_{ii'}$ by the $\mathbf{D}^{(i)}$, we have an analogous theorem for matrices; since the analogue of the condition $\sigma^2_{rr'} = \sigma^2_{ss'} = 1$ is the unitary property of the $\mathbf{D}^{(i)}$:

$$\mathbf{D}^{(i)\dagger}\,\mathbf{D}^{(j)} = \delta_{ij}\,\mathbf{1}$$

The implementation of the generalised MO method follows the same general lines as the method of section 12.3. The matrix $\mathbf{M}$ must carry a marker for those orbitals which are sent into linear combinations of the AO's and the actual linear coefficients kept in another matrix or series of smaller matrices. A full derivation of the theorem sketched in this section together with details of implementation are given in the original paper by M. Elder.

12.5 SYMMETRY ORBITALS AND THE MO METHOD

There are equations of the form (12.1.1) and (12.1.2) for each symmetry operation of the molecular symmetry group. For the moment we concentrate attention on just one operation $\hat{G}$, say. The operation $\hat{G}$ is chosen to leave the molecular Hamiltonian operator invariant and so the corresponding linear transformation matrix $\mathbf{D}$ leaves the matrix analogue, $\mathbf{H}^F$, invariant.

$$\mathbf{D}^\dagger\,\mathbf{H}^F\,\mathbf{D} = \mathbf{H}^F \qquad (12.5.1)$$

Multiplying on the left by $\mathbf{D}$ and using the unitary property of $\mathbf{D}$ gives

$$\mathbf{D}\,\mathbf{D}^\dagger\,\mathbf{H}^F\,\mathbf{D} = \mathbf{H}^F\,\mathbf{D} = \mathbf{D}\,\mathbf{H}^F$$

or

$$\mathbf{H}^F\,\mathbf{D} - \mathbf{D}\,\mathbf{H}^F = \mathbf{0} \qquad (12.5.2)$$

The RHF matrix *commutes* with the matrix representation of the symmetry operator. Now the matrix $\mathbb{D}$ can be diagonalised giving eigenvectors which are sets of coefficients defining *symmetry adapted* orbitals: orbitals which are simply multiplied by a constant under the operation of $\hat{G}$. If $Q^{(\alpha)}$ and $Q^{(\beta)}$ are two such eigenvectors, eigenvalues d_α and d_β respectively, then

$$\mathbb{D}\, H^F\, Q^{(\alpha)} = H^F\, \mathbb{D}\, Q^{(\alpha)} = d_\alpha\, H^F\, Q^{(\alpha)} \qquad (12.5.3)$$

and

$$\mathbb{D}\, H^F\, Q^{(\beta)} = H^F\, \mathbb{D}\, Q^{(\beta)} = d_\beta\, H^F\, Q^{(\beta)} \qquad (12.5.4)$$

Multiplying (12.5.3) on the left by $Q^{(\alpha)\dagger}$ and (12.5.4) by $Q^{(\beta)\dagger}$ and substituting the results gives

$$Q^{(\alpha)\dagger}(\mathbb{D}\, H^F - H^F\, \mathbb{D})Q^{(\beta)} = (d_\alpha - d_\beta)Q^{(\alpha)\dagger}\; H\, Q^{(\beta)}$$

since H^F and $\mathbb{D}$ are Hermitian. Using (12.5.2) we have

$$(d_\alpha - d_\beta)Q^{(\alpha)\dagger}\; H^F\, Q^{(\beta)} = 0 \qquad (12.5.5)$$

Thus, using a basis of symmetry adapted orbitals, the RHF matrix only has elements which "connect" orbitals of the same symmetry type ($d_\alpha = d_\beta$). Using this basis, the RHF matrix can, by permutations of rows and columns, be transformed into a "block form" with each block representing orbitals of a given symmetry type with respect to $\mathbb{D}$ and *all other elements zero*. If $\mathbb{Q}$ is the matrix of eigenvectors of $\mathbb{D}$ then the transformed RHF matrix

$$\mathbb{Q}^\dagger\, H^F\, \mathbb{Q}$$

has this block form.

In practice we have a *group* of $\mathbb{D}$'s, not a single element, and the $\mathbb{D}$'s, although each commutes with H^F, do not necessarily commute with each other and so simultaneous diagonalisation is not always possible. In these cases the generation of symmetry orbitals is still possible but the full derivations are outside the scope of this work, requiring the full apparatus of formal group representation theory. Fortunately, the generation of a set of symmetry orbitals requires only the knowledge of the irreducible representation matrices for the molecular symmetry group and a single recipe. The irreducible representation matrices of all molecular symmetry groups are well known and various tabulations exist. We can distinguish the different possible symmetry classifications by a label α and, if the irreducible representation matrices (in some standard axis system) are $D^{(\alpha)}(\hat{G})$, then the application of the operator

$$\rho_{ij}^{(\alpha)} = \sum_{\hat{G}} D^{\alpha}(\hat{G})_{ij} \hat{G} \tag{12.5.6}$$

to any atomic orbital produces a symmetry orbital of species α (or zero). Here, the symmetry operations themselves are denoted by $\hat{G}$ and the summation extends over all such operations in the molecular symmetry group. The representation matrices $D^{(\alpha)}(\hat{G})$ depend on the symmetry operation and on the symmetry species α. If the dimension of the representation matrices is g_α, for a particular choice of α, then multiplication of (12.5.6) by g_α/g (where g is the total number of symmetry operations in the group) produces a normalised operator: one which generates conventionally "normalised" symmetry orbitals. In fact, of course, the symmetry orbital's normalisation depends on the particular size of the AO overlap matrix elements.

The linear BeH_2 molecule provides a particularly simple example since all the $D^{(\alpha)}(\hat{G})$ are one dimensional. There are four possible symmetry types A_1, A_2, B_1 and B_2, in the usual nomenclature, and the irreducible representation matrices can be collected in the following table.

$\hat{G}$ / α	E	C_2^x	σ^{xy}	σ^{zx}
A_1	1	1	1	1
A_2	1	1	-1	-1
B_1	1	-1	1	-1
B_2	1	-1	-1	1

So, for example, $D^{(A1)}(C_2^{\ x}) = 1$ and $D^{(B2)}(E) = 1$ etc. The symmetry orbitals are obtained by applying (12.5.6) for our particular choice of α and $\hat{G}$. The possible symmetry orbitals arising from the hydrogen 1s AO's can be obtained by applying the relevant $\rho_{ij}^{(\alpha)}$ to ϕ_1 and are given below.

A_1 symmetry: $\phi_1 + C_2^{\ x}\phi_1 + \sigma^{xy}\phi_1 + \sigma^{zx}\phi_1$

$$= \phi_1 + \phi_2 + \phi_2 + \phi_1 = 2(\phi_1 + \phi_2)$$

or, when normalised,

$$\frac{1}{(2 + 2S_{12})^{\frac{1}{2}}}(\phi_1 + \phi_2).$$

A_2 symmetry: $\phi_1 + C_2^{\ x}\phi_1 - \sigma^{xy}\phi_1 - \sigma^{zx}\phi_1$

$$= 0,$$

there are *no* linear combinations of hydrogen 1s orbitals of A_2 symmetry. Similarly, there is no B_2 symmetry orbital.

B_1 symmetry: $\phi_1 - C_2^{\ x}\phi_1 - \sigma^{xy}\phi_1 + \sigma^{zx}\phi_1$

$$= 2(\phi_1 - \phi_2)$$

which, when normalised, gives

$$\frac{1}{(2 - 2S_{12})^{\frac{1}{2}}}(\phi_1 - \phi_2)$$

Similar reasoning shows that the Be 1s and 2s orbitals *are* symmetry orbitals of species A_1 and the 2p orbital is a symmetry orbital of B_1 type. We can collect these coefficients into a matrix

$$\mathbb{Q} = \begin{pmatrix} Q_{11} & 0 & 0 & Q_{41} & 0 \\ Q_{12} & 0 & 0 & Q_{42} & 0 \\ 0 & 1 & 0 & 0 & 0 \\ 0 & 0 & 1 & 0 & 0 \\ 0 & 0 & 0 & 0 & 1 \end{pmatrix}$$

where $Q_{11} = Q_{12} = (2 + 2S_{12})^{-\frac{1}{2}}$ and $Q_{41} = -Q_{42} = (2 - 2S_{12})^{-\frac{1}{2}}$. The columns of Q have been chosen to define symmetry orbitals and to have symmetry orbitals of the same type adjacent. Thus, as can be verified using our computed matrix elements,

$$\mathsf{Q}^{\dagger}\,\mathsf{S}\,\mathsf{Q} \qquad \mathsf{Q}^{\dagger}\,\mathsf{H}\,\mathsf{Q} \qquad \text{and} \qquad \mathsf{Q}\,\mathsf{H}^{F}\,\mathsf{Q}$$

are all in block form, consisting of a 3×3 block of A_1 symmetry and a 2×2 block of B_1 symmetry and zeroes elsewhere.

This result has obvious application in the iterative solution of the RHF equation; the eigenvalues of the 5×5 matrix can be had by solving two separate, smaller, matrices: in our case a 2×2 and a 3×3. In the general case, the symmetry orbital transformation can be applied at the same time as the orthogonalisation of H^{F}. The matrix

$$\mathsf{X} = \mathsf{Q}(\mathsf{Q}^{\dagger}\,\mathsf{S}\,\mathsf{Q})^{-\frac{1}{2}} \qquad (12.5.7)$$

can be used in place of the orthogonalisation matrix $\mathsf{S}^{-\frac{1}{2}}$ in the iterative solution of the RHF equation and we obtain, in one step, a transformation to a set of normalised, orthogonal symmetry orbitals. If the full m dimensional H^{F} matrix is broken into a set of m_α dimensional matrices then the computation of the eigenvectors of H^{F} is speeded up considerably since the diagonalisation process consumes time proportional to m^3 or m^4, and

$$m^3 > \sum_{\alpha} m_\alpha^3 \qquad \left(\sum_{\alpha} m_\alpha = m\right).$$

12.6 THE USE OF SYMMETRY ORBITALS IN STORAGE COMPRESSION

If we write down the expression for the one-electron integrals over symmetry orbitals in terms of the AO one-electron integrals the general result, using (12.5.6), is

$$\int dr\, \gamma_r^{(\alpha)}\, \hat{h}\, \gamma_s^{\alpha} = \int dr (\rho_{ij}^{(\alpha)} \phi_k)\, \hat{h}\, \rho_{ij}^{(\alpha)} \phi_\ell \qquad (12.6.1)$$

where each symmetry orbital $\gamma_r^{(\alpha)}$ is obtained from an AO by the action of one of the operators $\rho_{ij}^{(\alpha)}$. In particular, the integral involving the A_1 orbital of BeH_2 derived in the last section is, apart from normalisation,

$$\int dr(\phi_1 + \phi_2)\hat{h}(\phi_1 + \phi_2)$$

$$= \int dr\phi_1\hat{h}\phi_1 + 2\int dr\phi_1\hat{h}\phi_2 + \int dr\phi_2\hat{h}\phi_2$$

and, since ϕ_1 and ϕ_2 are equivalent in the molecule, the integral is simply

$$2\left(\int dr\phi_1\hat{h}\phi_1 + \int dr\phi_1\hat{h}\phi_2\right)$$

That is, AO integrals which involve orbitals related by the operations of the symmetry group *contribute equally* to the corresponding symmetry orbital integral. The conditions that this result should be true in general are

i) the symmetry group must have one-dimensional irreducible representation matrices;

and

ii) the integrand should be invariant with respect to the operations of the molecular symmetry group - in our case the operator $\hat{h}$ should be totally symmetric with respect to these operations.

The proof of the general case involves the expansion of (12.6.1) and some of the elementary properties of the $\rho_{ij}^{(\alpha)}$ operators. When the molecular symmetry group has two or three-dimensional irreducible representations a simple device enables the conclusions obtained for the one-dimensional case to be used: this extension has been given by R.M. Pitzer.

Now, if AO integrals contribute equally to symmetry orbital integrals we can simply compute *one* of the AO integrals, multiply it by the number of symmetry - related integrals, and, *provided we transform to symmetry orbitals* (via (12.5.7)) we have the correct one-electron hamiltonian matrix. Again, in the MO calculation, we can use the one-electron symmetry of the matrix $\mathsf{G}(\mathsf{R})$ and a simple extension of our result shows that, if we form a partial G matrix which has each *symmetry distinct* two-electron integral multiplied by the number of symmetry related integrals, transformation of this G matrix to a symmetry orbital basis gives the correct full matrix $\mathsf{G}(\mathsf{R})$. There is one cautionary point to be made about this symmetry orbital method. The symmetry orbital matrix should have zeroes in the positions corresponding to integrals of different symmetry types. These zeroes occur by cancellation of equal magnitudes of opposite sign during the transformation to a symmetry orbital basis. The scaling method outlined above will *not have the correct cancellation properties* and so, although the elements within the symmetry blocks are correct, the blocks of zeroes will *contain rubbish* which must be discarded. Thus, for example, during the iterative solution of the RHF equation we must arrange to diagonalise the small symmetry blocks only, or, if we diagonalise the whole matrix, the inter-symmetry elements must be set to zero.

12.7 SUMMARY

We have covered a fairly large amount of ground in this Chapter and mentioned a number of alternative procedures for using molecular symmetry; therefore a summary of the main results obtained is not out of place.

i) From the point of view of *processing time only* it is always useful to spot equivalences and zeroes among the molecular integrals. When M is a simple permutation matrix this saving is particularly easy to implement.

ii) When M contains linear combinations then, for an integral file for general use, the compromise of using only the "permutation symmetries" gives a useful time saving.

iii) In the special case of the LCAOMO method, the one-electron symmetry of H^F enables considerable savings of time and storage space to be made by computing *and storing* only the symmetry unique integrals.

iv) The implementation of the LCAO MO method can be speeded up considerably by using a basis of symmetry orbitals since the symmetry unique contributions to $\mathbf{G}(\mathbf{R})$ need only be multiplied by a scaling factor to obtain the correct matrix.

v) Savings of storage space are also savings of processor time since, at each iteration of the MO method, the electron repulsion integral file must be processed.

SUGGESTIONS FOR FURTHER READING

The use of the permutation matrix was first developed by the POLYATOM team: see Appendix B.

"On the Use of Symmetry in SCF Calculations" by P.D. Dacre in Chem.Phys.Letters, 7, 47 (1970) describes the storage-compression algorithm used in 12.3.

"On the Use of Symmetry in Molecular SCF Calculations" by M. Elder in Int.J.Quant.Chem., 6, 75 (1973) generalises Dacre's algorithm to include all symmetry operations.

The use of symmetry orbitals to speed up molecular integral handling is described in "The Contribution of AO Integrals to Symmetry Orbital Integrals" by R.M. Pitzer in J.Chem.Phys., (in press).

The definition, use and formal properties of the operators $\rho_{ij}^{(\alpha)}$ defining symmetry orbitals in given in "Symmetry" by R. McWeeny (Pergamon Press 1963).

13 LOCALISED DESCRIPTIONS OF ELECTRONIC STRUCTURE

13.1 CHEMICAL CONSEQUENCES OF INVARIANCE

In Chapter 9 we obtained results concerning the invariance of the one-determinant LCAOMO wave function with respect to two kinds of linear transformation. We used the fact that the MO calculation is invariant against linear transformations among the basis orbitals as a manipulative aid to solve the RHF equation in a non-orthogonal AO basis. In this Chapter we will see how it is possible to gain some insight into the nature of the valence electronic structure by using these invariance properties. Each of the two types of invariance outlined in Chapter 9 - invariance with respect to transformations among the occupied MO's and among the AO's - have important applications. The MO invariance gives a chemical interpretation of the molecular electronic distribution and the AO invariance suggests a natural extension of the MO method and a "rehabilitation" of the VB method in a restricted form.

13.2 LOCALISED MOLECULAR ORBITALS AND THE TWO ELECTRON BOND

As we proved in section 9.3, the total molecular wave function (in the one-configuration approximation) is not changed by a unitary transformation among the occupied orbitals. Thus the AO density matrix, which summarises the electron density, is not changed by such transformations. This degree of arbitrariness in the definition of the occupied orbitals suggests that we could use the implied freedom to choose a unitary transformation among the occupied MO's to yield a set of MO's which are particularly adapted to chemical interpretation. This transformation would form a useful bridge from molecular orbitals to the familiar concepts of empirical chemistry. If we recall the original discussion of the qualitative forms of the MO's in Chapter 5, it is clear that the molecular orbitals arising "naturally" as the solutions of the RHF equation are *de-localised*: a typical MO will have significant contributions (non-zero elements of T) from all the AO's of the molecule. This description of the molecular electron density is in marked contrast to the familiar chemical idea of localised electron-pair bonds and the invariance of bonds, lone pairs and inner shells with respect to changes in molecular environment. A contour diagram of the valence electron density in a molecule *does* reveal regions of high electron density in the bond and lone pair regions but this is the result of the *superposition* of the essentially de-localised contributions from the individual MO's. The electron density in the C-H bond region is computed to be similar in methane and ethane, for example. Thus the unobservable individual MO contributions do add up to form a chemical picture. The question is: can we "draw new lines" around the local bond regions and define *localised molecular orbitals?* If this can be done the total electron density can be viewed as the addition of *spatially distinct* regions of high electron density rather than the superposition of *spatially overlapping* densities.

It is desirable to have some well-defined criteria for the computation of localised MO's from the RHF de-localised MO's. One of the first, and most successful, methods is best introduced by writing the expression for the total energy of a one-configuration wave function in terms of integrals involving the molecular orbitals. The relevant equation is (5.2.13). The MO density matrix has a particularly simple form, having 2 on the main diagonal and zeroes elsewhere: $R_{ij} = \delta_{ij}$.

$$\mathsf{P}_{MO} = 2\,\mathsf{R}_{MO} = 2\,\mathbf{1} \tag{13.2.1}$$

Using this simple form for P, the energy expansion becomes

$$E = 2\sum_{i=1}^{n/2} \int dr\psi_i(r)\hat{h}\psi_i(r)$$

$$+ \sum_{i,j=1}^{n/2} \{2(\psi_i\psi_i,\psi_j\psi_j) - (\psi_i\psi_j,\psi_i\psi_j)\} \tag{13.2.2}$$

in which the only MO electron repulsion integrals appearing are the two types $(\psi_i\psi_i,\psi_j\psi_j)$ and $(\psi_i\psi_j,\psi_i\psi_j)$. The physical interpretation of the first of these integrals is obvious; it is the "classical" repulsion between an electron in MO ψ_i and one in MO ψ_j. The other integral - the exchange term - has no classical analogue and is a consequence of the anti-symmetry requirement. It is often usual to interpret chemical effects in terms of a semi-classical electrostatic model and so we give the following prescription for forming localised molecular orbitals:

> The Localised Molecular Orbitals (LMO's) for an electronic system are those for which the energy expression has the form (13.2.2) with minimum contribution from the exchange integrals.

That is, we ask for a set of orbitals ψ_{Li} such that

$$E = 2\sum_{i=1}^{n/2} \int dr\psi_{Li}(r)\hat{h}\psi_{Li}(r)$$

$$+ \sum_{i,j=1}^{n/2} (\psi_{Li}\psi_{Li},\psi_{Lj}\psi_{Lj})(2 - \delta_{ij}) \tag{13.2.3}$$

The exchange integrals are two-label quantities and so can be written as a matrix K with elements

$$K_{ij} = (\psi_i\psi_j,\psi_i\psi_j)$$

Here, the diagonal elements are the "self-repulsion" coulomb terms. The unitary transformation which *diagonalises* the symmetric matrix K defines a set of orbitals for which the exchange integrals are small. That is, the matrix L such that

$$\mathsf{L}^{\dagger}\,\mathsf{K}\,\mathsf{L} \;=\; \text{diagonal matrix}$$

has columns which express the ψ_{Li} in terms of the original delocalised MO's ψ_i. Collecting the equations which connect the AO's, MO's and LMO's we have

$$\boldsymbol{\psi} = \boldsymbol{\varphi}\,\mathsf{T}$$

and

$$\boldsymbol{\psi}_L = \boldsymbol{\psi}\,\mathsf{L}$$

thus,

$$\boldsymbol{\psi}_L = \boldsymbol{\varphi}\,\mathsf{T}\,\mathsf{L}$$

gives the definition of the LMO's in terms of the AO's. (Note that T is m×n/2; L is n/2×n/2 therefore $\mathsf{T}\,\mathsf{L}$ is m×n/2, defining m/2 LMO's in terms of m AO's). This particular method of choosing localised orbitals has the advantage of requiring no new programming techniques, being a simple diagonalisation. A more complete procedure would be to minimise the exchange contributions and the "off diagonal" coulomb repulsion while simultaneously maximising the self-repulsion terms. There are technical difficulties associated with this maximum/minimum problem which we discuss in section 13.4.

The results of applying our simple localisation procedure to the MO's of BeH_2 are illustrated below (Fig. 13.1 and Table 13.1). The matrix K, obtained from the AO repulsion integrals and T via equation (9.5.1), is given together with the matrix L which diagonalises K. The product $\mathsf{T}\,\mathsf{L}$ defining the LMO's in terms of the AO's is listed and, finally, a contour diagram of the LMO's defined by $\mathsf{T}\,\mathsf{L}$ is given. As the table shows, the inner shell 1s orbital of the Be atom is essentially unchanged (the relevant K_{ij} are very small). More important, the Be-H "bond orbital" does have a chemically appealing

Table 13.1 Localised MO's for BeH_2

K	0.0	0.0206	0.0026
	0.0206	0.0	0.1472
	0.0026	0.1472	0.0
L	0.9999	-0.0066	-0.0066
	0.0110	0.7326	0.6806
	0.0022	-0.6806	0.7326
TL	-0.0020	0.6229	0.0421
	-0.0020	0.0421	0.6229
	0.9869	-0.1555	-0.1555
	0.0545	0.2963	0.2963
	0.0	-0.3119	0.3119

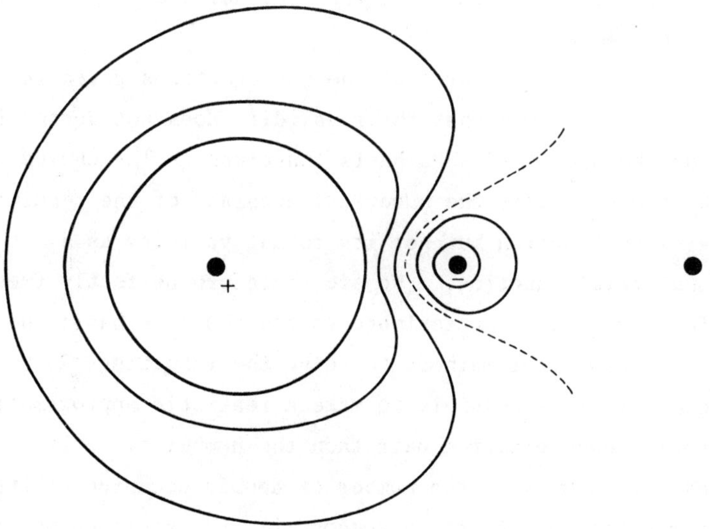

Figure 13.1 The valence LMO for Be-H

localised character in line with chemical intuition.

The localisation process can be applied to homologous series of molecules and the localised orbitals describing similar bonds can be examined to obtain insight into the effect of environment. There are a variety of criteria used to define LMO's and other methods are referenced at the end of this Chapter.

13.3 THE DIRECT CALCULATION OF LMO's

The considerations of the last section encourage the view that the localised molecular orbitals are, for the chemist, more fundamental elements of molecular structure than the de-localised MO's. In fact, we could say that the calculation of de-localised MO's is just a numerical step in the computation of LMO's. If we prefer to work with LMO's it is natural to ask if these localised orbitals can be computed *directly* without the intermediate step of calculating the de-localised MO's - i.e. by-passing the LCAO method.

Examination of the forms of the RHF equations given in Chapters 4 and 11 shows that their validity does not depend in any way on the use of AO's as basis functions. The choice of basis functions effects the numerical accuracy of the resulting molecular wave function but not its formal validity as a variational trial function. We are therefore perfectly free to choose functions which approximate to the LMO's as basis and use our existing numerical methods to solve the resulting "RHF" equation. If it is possible to make a realistic approximation to an LMO for each electron pair then the number of basis functions will simply be the number of doubly occupied orbitals.

Restricting our attention temporarily to saturated molecules, we can make a good guess at the spatial form of a single electron-pair σ bond based on general chemical knowledge and a study of the forms of the LMO's obtained from LCAOMO's. Such a bond is expected to be cylindrically symmetrical and roughly elliptical in shape. A function having this spatial form is given by

$$\exp(-\alpha x_L^2 - \alpha y_L^2 - \beta z_L^2) \qquad (13.3.1)$$

where α and β are measures of the relative radii of the ellipsoid which is centred at (x_L, y_L, z_L) along the internuclear axis. The "Gaussian" form of the function (13.3.1) was chosen in view of our previous experience with the computational advantages of GTF's ($\alpha=\beta$ defines a 1s GTF). The use of n/2 functions of this type to describe the spatial distribution of n electrons gives an important simplification of the MO equations. Clearly, in a basis of n/2 orthogonalised functions the $\bar{\mathsf{R}}$ matrix has "MO form": $R_{ij} = \delta_{ij}$. Therefore the R matrix in the non-orthogonal basis (13.3.1) is obtainable directly from (9.2.5) and (9.4.3)

$$\mathsf{R} = \mathsf{S}^{-\frac{1}{2}}\,\bar{\mathsf{R}}\,\mathsf{S}^{-\frac{1}{2}}$$

$$= \mathsf{S}^{-\frac{1}{2}}\,\mathsf{1}\,\mathsf{S}^{-\frac{1}{2}} = (\mathsf{S}^{-\frac{1}{2}})^2$$

$$\text{i.e.} \qquad \mathsf{R} = \mathsf{S}^{-1}. \qquad (13.3.2)$$

The energy expression for the LMO one-determinant molecular wave function is then

$$E = 2\,\mathrm{tr}\,\mathsf{H}\,\mathsf{S}^{-1} + \mathrm{tr}\,\mathsf{G}(\mathsf{S}^{-1})\mathsf{S}^{-1} \qquad (13.3.3)$$

and *no iterative calculation is involved.* Apparently there is no freedom for the variation principle to optimise the approximate LMO wave function. In fact the variation principle can be applied to (13.3.3) by minimising E with respect to the elements of H, S and the $(ij,k\ell)$ in G: that is, with respect to *the basis functions* which define these molecular integrals. Use of the variation principle to optimise (13.3.3) with respect to the parameters contained in (13.3.1) - α, β and the point of origin L - will give the best possible approximate LMO description of the molecule consistent with the form (13.3.1) of the LMO's. Unfortunately the dependence of the H_{ij}, S_{ij} and the repulsion integrals on these parameters is non-linear and the optimisation process cannot be reduced to the pseudo-eigenvalue problem of the RHF method. In principle, we have to compute the total energy, that is all the molecular integrals, for a wide range of values of α, β and L for each LMO until a minimum is found.

The problem becomes, computationally, much more tractable if the ellipsoidal form of the functions (13.3.1) is replaced by a spherical 1s GTF form

$$\exp(-\alpha r_L^{\,2})$$

but at the expense of using a spherical function to approximate an LMO which we have good reason to suspect is non-spherical. However, for qualitative and semi-quantitative purposes, the use of a simple spherical approximation to each LMO in a molecule has proved to be very useful. For computational purposes it is useful to re-cast equation (13.3.3) into a form where the summations are over the molecular integral indices directly, since in practice we compute the distinct molecular integrals and form the contribution to E from each one. Using the definition of $\mathsf{G}(\mathsf{R})$ and a temporary notation T for S^{-1}, we find

$$E = 2\sum_{i,j=1}^{n/2} T_{ij}H_{ji} + \sum_{i,j,k,\ell=1}^{n/2} (ij,k\ell)[2T_{ij}T_{k\ell} - T_{ik}T_{j\ell}] \qquad (13.3.4)$$

Our ordering convention for the labels of the one-electron integrals and the repulsion integrals enables the summation to be re-cast as

$$\begin{aligned} E = 2 &\sum_{i,j} (2 - \delta_{ij})T_{ij}H_{ji} \\ &+ \sum_{i,j,k,\ell} (ij,k\ell)[2T_{ij}T_{k\ell} - (T_{ik}T_{j\ell} + T_{jk}T_{i\ell})] \\ &\times 2^{3-\delta_{ij}-\delta_{k\ell}-\varepsilon} \end{aligned} \qquad (13.3.5)$$

where δ_{ij} is the Kroneker delta and ε is unity if $[ij]=[k\ell]$ and zero otherwise. In (13.3.5) the summations range over $i \geqslant j$; $k \geqslant \ell$; $[ij] \geqslant [k\ell]$ only. Given programs for the computation of the H_{ij} and $(ij,k\ell)$, this expression is easily programmed and gives an approximation to the total electronic energy for the particular choice of α's and L's. The total *molecular* energy is given by (13.3.5) plus the nuclear repulsion energy

$$\sum_{\alpha>\beta=1}^{N} \frac{Z_\alpha Z_\beta}{R_{\alpha\beta}}$$

which, if the nuclear geometry is fixed, is a constant and may be omitted from the calculation. Since the molecular integrals are used as they are computed it is, strictly, not necessary to store them. In practice, a systematic variation of the α and L parameters will be carried through and it is often convenient to store the integrals and only re-compute the new ones when an orbital parameter is changed. In order to minimise (13.3.5) we need an efficient non-linear minimisation procedure which does not use derivatives since analytical derivatives of (13.3.4) with respect to the α's and L's are too cumbersome to compute. Non-linear optimisation is a rapidly developing field with new methods being developed in econometrics, engineering and science. The most commonly used methods in quantum chemistry are due to Fletcher, Reeves and Powell*. The results of applying this direct LMO calculation for BeH_2 are summarised in Table 13.2 and the orthogonal Be-H LMO is plotted in Fig. 13.2 for comparison with the LMO's obtained earlier from the LCAOMO's.

13.4 OPTIMUM BOND HYBRIDS

The method of localised molecular orbitals outlined in the previous sections has important conceptual advantages over the simple LCAOMO scheme, but since the total wave function is invariant against the localisation procedure, there is no numerical value in the process. It is, from a purists' point

---

* FORTRAN implementations of many of these methods are obtainable from the SRC ATLAS Computing Laboratory at Chilton, Berkshire.

Table 13.2 Single GTF per electron-pair results for BeH_2

Description	Distance of centre from Be atom	Optimum α
Be-H bond pair	2.0044	0.2423
Be $1s^2$ lone pair	0.0	3.8581

Total Electronic Energy = -16.60806 a.u.

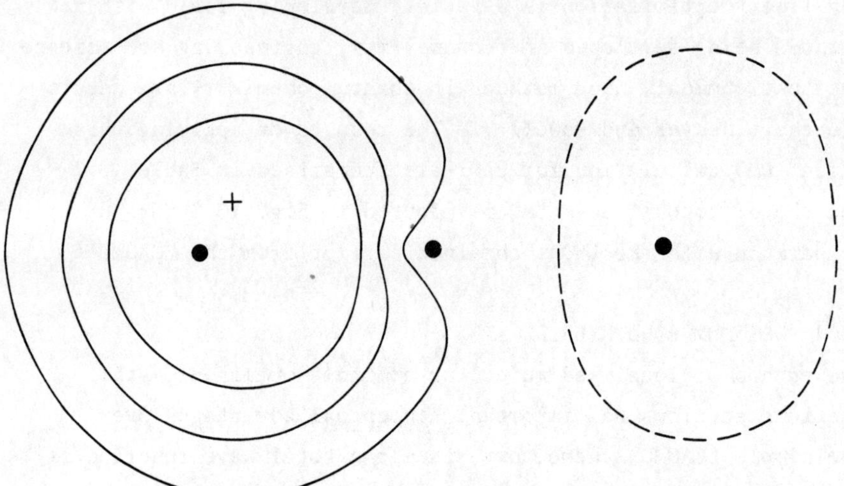

Figure 13.2 The Be-H LMO computed directly using a single GTF for each electron pair

of view, arbitrary. However, we can pose the localisation problem in a different way by using the invariance of the MO wave function against transformations among the AO functions. This localised picture can be of conceptual *and* numerical value beyond the one-configuration model.

It is probably felt by most chemists that it is the *constituents* of the LMO's, rather than the localised orbitals themselves, which are the invariant elements of molecular electronic structure - the hybrid atomic orbitals. Two LMO's, which are slightly different due to the different molecular environments of two chemically similar electron pairs, may both be regarded as being formed from linear combinations of the same two hybrid AO's in different amounts. In this view the small differences in the LMO's due to environmental factors are taken up in the linear coefficients which define the LMO in terms of the hybrids. It is therefore chemically meaningful to ask:

> What is the hybrid AO basis which gives the best possible pairwise bonding scheme?

(again, restricting our attention to saturated systems). That is; in what AO basis do the LMO's appear as predominantly linear combinations of pairs of hybrids? Mathematically, can we find a matrix A such that

$$\boldsymbol{\varphi}_A = \boldsymbol{\varphi}\mathsf{A} \qquad (13.4.1)$$

is a hybrid AO basis satisfying the condition that

$$\mathsf{R}_A = \mathsf{A}^{-1}\,\mathsf{R}\,\mathsf{A}^{-1\dagger} \qquad (13.4.2)$$

is as close as possible to a set of 2×2 sub-matrices along the principle diagonal of R_A with zeroes elsewhere. If R_A has this form then the LMO's have the required limited form. Strictly, in order to maintain an inter-atomic valence interpretation of the LMO's as linear combinations of atomic hybrid orbitals, the matrix A should be constrained to "mix" only AO's on the same atom - A should have atomic block form.

The calculation of A is, then, a well-defined optimisation problem. A good approximation to A and, therefore, a useful starting point in the numerical optimisation is the matrix composed of the conventional hybridisation coefficients relevant to the molecule: sp hybrids for BeH_2, sp^3 for CH_4, etc. The effect of transformation (13.4.2) on the R matrix for BeH_2 when

$$\mathsf{A} = \begin{pmatrix} 1.00 & 0 & 0 & 0 & 0 \\ 0 & 0 & 1.00 & 0 & 0 \\ 0 & 0 & 0 & 0 & 0 \\ 0 & 0.707 & 0 & 0.707 & 0 \\ 0 & -0.707 & 0 & 0.707 & 1.00 \end{pmatrix} \qquad (13.4.3)$$

is

$$\mathsf{R}_A = \begin{pmatrix} 0.39 & 0.27 & 0.02 & -0.01 & -0.09 \\ 0.27 & 0.19 & -0.01 & 0.02 & -0.02 \\ 0.02 & -0.01 & 0.39 & 0.27 & -0.09 \\ -0.01 & 0.02 & 0.27 & 0.19 & -0.02 \\ -0.09 & -0.02 & -0.09 & -0.02 & 1.02 \end{pmatrix} \qquad (13.4.4)$$

Thus, simple chemical hybridisation goes a long way towards a solution of (13.4.2) for molecules of conventional structure. Of course, for molecules which cannot be assigned a unique electron pair bonding scheme - aromatics, boranes - this simple procedure is not applicable.

We may use, in place of (13.4.3) which defines conventional hybrid AO's, the matrix $\mathsf{W}\,\mathsf{V}\,\mathsf{S}_h^{-\frac{1}{2}}$ discussed in Chapter 9 and obtain the matrix of charges and bond orders over an *orthogonal* set of hybrid orbitals. For BeH_2

$$\mathsf{W}\,\mathsf{V}\,\mathsf{S}_h^{-\frac{1}{2}} = \begin{pmatrix} 1.579 & -0.216 & 0.0 & -0.826 & 0.172 \\ -0.216 & -1.579 & 0.0 & 0.172 & -0.826 \\ -0.016 & -0.026 & 1.0 & -0.168 & -0.168 \\ -0.472 & -0.472 & 0.0 & 1.030 & 1.030 \\ 0.703 & -0.703 & 0.0 & -1.200 & 1.200 \end{pmatrix} \qquad (13.4.5)$$

$$\mathsf{R}_A = \begin{pmatrix} 0.56 & 0.49 & 0.04 & 0.00 & -0.01 \\ 0.49 & 0.44 & 0.00 & -0.04 & 0.01 \\ 0.04 & 0.00 & 0.56 & 0.49 & -0.01 \\ 0.00 & -0.04 & 0.49 & 0.44 & 0.01 \\ -0.01 & 0.01 & -0.01 & 0.01 & 1.00 \end{pmatrix} \qquad (13.4.6)$$

The elements of R_A between orbitals which are not formally bonded are again small, and (13.4.6) has the convenient property that $2\mathrm{tr}\mathsf{R}_A = n$. In particular, the sum of the diagonal elements of each bond orbital pair is close to unity, showing clearly the "two electrons per bond" property.

Having defined R_A over an orthogonal basis of hybrids, we can now choose any further improvements to the matrix A to be *unitary*. If a better approximation to A is given by $\mathsf{A}\,\mathsf{A}'$ then

$$\mathsf{A}'^{\dagger} = \mathsf{A}^{-1}$$

and (13.4.2) becomes

$$\mathsf{R}_{A'} = \mathsf{A}'^{\dagger}\,\mathsf{R}_A\,\mathsf{A}' \qquad (13.4.7)$$

This equation has some formal similarity to the definition of the eigenvalues and eigenvectors of a matrix

$$\boldsymbol{\epsilon} = \mathsf{U}^{\dagger}\,\mathsf{H}^{F}\,\mathsf{U} \qquad (13.4.8)$$

and this similarity suggests a method of computing an optimum A'. By analogy with the Jacobi method for the solution of (13.4.8) - discussed in Chapter 7 - we define A' as the product of a series of 2×2 unitary transformations each of the form

$$\begin{pmatrix} c & -s \\ s & c \end{pmatrix} \qquad (c=\cos\theta,\ s=\sin\theta)$$

but the condition for the solution of (13.4.7) is rather more complex than for (13.4.8). The diagonal elements and *certain* off-diagonal elements of $\mathsf{R}_{A'}$ are to be maximised at the expense of minimising the other off-diagonal elements of $\mathsf{R}_{A'}$. The effect of a typical 2×2 transformation on the off-diagonal element of R_A is to transform it into

$$C\,R_{A\,ij} + S(R_{Ajj} - R_{Aii}) = R_{A'ij} \tag{13.4.9}$$

(where $C = \cos 2\theta$, $S = \sin 2\theta$)
We require that (13.4.8) be a maximum if i and j are bonded and a minimum if i and j are not bonded:

$$\frac{d}{d\theta}\,R_{A'ij} = 0$$

and

$$\frac{d^2 R_{A'ij}}{d\theta^2} > 0$$

in one case and <0 in the other. Hopefully, repeated transformations of this type reduce R_A to the required form. There are often convergence difficulties associated with this type of maximisation/minimisation problem. The solution lies on a "saddle point" on the surface coordinates.

Having, either through (13.4.7) or (13.4.5), defined a set of hybrid AO's we can use a model wave function for electron pair *of the VB type* since there are only three possible VB structures for each electron pair. Even if part of the electronic structure is essentially delocalised - aromatics - the partial localisation procedure enables a chemically meaningful use of the VB method to be made.

13.5 THE SEPARATE ELECTRON PAIR MODEL

If the ground-state wave function of a typical two electron "group" (bond, lone pair, inner shell) is Ψ^R, then the wave function describing the electronic structure of a molecule containing n paired electrons is

$$\Psi = \hat{A} \prod_{R=1}^{n/2} \Psi^R \tag{13.5.1}$$

and the total electronic energy of such a system is given by

$$E = \int d\tau\ \Psi\ \hat{H}\ \Psi / \int d\tau\ \Psi\Psi \qquad (13.5.2)$$

In general we may be dealing with wave functions of the R'th group which represent excited states of that electron pair, but for present purposes consideration of the ground-state function of each group is sufficient. The anti-symmetric function Ψ has a formal similarity to the one-configuration, determinantal, function (3.3.3) in that the one-electron orbitals have been replaced by electron pair functions (sometimes called "geminals"). This formal similarity is particularly useful if the functions Ψ^R satisfy the so-called *strong orthogonality* condition

$$\int dx_1\ \Psi^R(x_1,x_2)\Psi^S(x_1,x_2) = 0 \qquad (13.5.3)$$

if $R \neq S$. That is, if integration over only one set of variables is enough to ensure orthogonality. We will normally choose the individual Ψ^R to be of orbital form and anti-symmetric with respect to their own variables, in this case (13.5.3) can be satisfied by using one set of orbitals for building up Ψ^R and an orthogonal set for Ψ^S

$$\int dr\ \phi_i^R(r)\phi_j^S(r) = \delta_{RS} \qquad (13.5.4)$$

If the strong orthogonality condition is met and the orbitals *within* a group are orthogonal

$$\int dr\ \phi_i^R(r)\phi_j^R(r) = \delta_{ij}$$

then (13.5.2) takes a form exactly analogous to the Slater's rule expression for the energy of a one-configuration wave function:

$$E = \sum_{R=1}^{n/2} H^R + \sum_{R>S=1}^{n/2} (J^{RS} - K^{RS}) \qquad (13.5.5)$$

Here, H^R is the energy of the R'th group of electrons in the field of the nuclei and each other

$$H^R = \int dx_1 \int dx_2\ \Psi^R(x_1,x_2)\hat{h}^R\ \Psi^R(x_1,x_2) \tag{13.5.6}$$

where

$\hat{h}^R$ is the R-group Hamiltonian

$$\hat{h}^R = \hat{h}(1) + \hat{h}(2) + \hat{g}(1,2)$$

The coulomb and "exchange" interactions between the electrons of group R and those of the other groups are summarised in the J and K terms respectively

$$J^{RS} = \int d\tau\ \Psi^R(x_1,x_2)^2\left(\frac{1}{r_{13}} + \frac{1}{r_{14}} + \frac{1}{r_{23}} + \frac{1}{r_{24}}\right)\Psi^R(x_3,x_4)^2$$

$$(13.5.7)$$

$$K^{RS} = \int d\tau\ \Psi^R(x_1,x_2)\Psi^S(x_3,x_4)$$

$$\times \left|\frac{1}{r_{13}}\Psi^R(x_3,x_2)\Psi^S(x_1,x_4) + \frac{1}{r_{14}}\Psi^R(x_4,x_2)\Psi^S(x_3,x_1)\right.$$

$$\left.+ \frac{1}{r_{23}}\ \Psi^R(x_1,x_3)\Psi^S(x_2,x_4) + \frac{1}{r_{24}}\Psi^R(x_1,x_4)\Psi^S(x_3,x_2)\right|$$

where $d\tau = dx_1dx_2dx_3dx_4$. The inter-pair repulsion terms can be combined with the R-group energy to define an effective Hamiltonian for group R. We can write (13.5.5) in the form

$$E = \sum_{R=1}^{n/2} H^R_{eff} - \sum_{R>S=1}^{n/2} (J^{RS} - K^{RS}) \tag{13.5.8}$$

where

$$H^R_{eff} = H^R + \sum_{S\neq R=1}^{n/2} (J^{RS} - K^{RS}) \tag{13.5.9}$$

can be interpreted as the expectation value of some R-group Hamiltonian *operator* $\hat{H}^R_{eff}$ over the R-group wave function Ψ^R

$$H^R_{eff} = \int dx_1 \int dx_2\ \Psi^R\ \hat{H}^R_{eff}\ \Psi^R \tag{13.5.10}$$

It can be shown that the variational solution of (13.5.10) minimises (13.5.9): thus, by choosing normalised functions to minimise (13.5.9) for each electron pair, the total energy is minimised and the resulting total wave function is the best possible one of electron-pair form. The method of approaching the computation of electron-pair functions is precisely analogous to the earlier variational methods; we must choose a flexible form for each Ψ^R and optimise the parameters in the function.

An obvious orbital form for the trial function Ψ^R for each electron pair is the VB function - discussed in 5.3 for a single electron pair - representing a mixture of covalent and ionic forms for each group,

$$\Psi^R = a\,\Psi^R_{HL} + b\,\Psi^R_{+-} + c\,\Psi^R_{-+} \qquad (13.5.11)$$

in the notation of (5.3.4). In this case the minimisation of (13.5.9) corresponds to the diagonalisation of a 3×3 matrix to obtain optimum a, b, c for each electron pair. Since the Hamiltonian $\hat{H}^R_{eff}$ contains the electron repulsion between group R and the other two-electron pairs, the minimisation of all n/2 equations (13.5.9) is iterative: an initial form for the structure of each group is iteratively refined by repeated minimisation of (13.5.9) in turn until self-consistency is reached. The expressions for the matrix elements of $\hat{H}^R_{eff}$ over the functions of (13.5.11) can be obtained by the use of Slater's rules.

Each two electron group involves the use of two hybrid orbitals, so that the charge and bond-order matrix P^R for each group is 2×2 and contains two degrees of freedom;

the group bond order $X_R = P^R_{12} = P^R_{21}$

and the group "polarity" $Y_R = \frac{1}{2}(P^R_{22} - P^R_{11})$.

In terms of the solutions a, b and c of (13.5.11) each P^R is given by

$$a^2\begin{pmatrix}1 & 0\\0 & 1\end{pmatrix} + b^2\begin{pmatrix}0 & 0\\0 & 2\end{pmatrix} + c^2\begin{pmatrix}2 & 0\\0 & 0\end{pmatrix} \qquad (13.5.12)$$

For convenience of notation, we re-name the functions Ψ^R_{HL}, Ψ^R_{+-} and Ψ^R_{-+} as Ψ^R_1, Ψ^R_2 and Ψ^R_3 respectively:

$$\Psi^R_1 = \frac{1}{\sqrt{2}}(\det\{\phi^R_1\ \bar{\phi}^R_2\} - \det\{\bar{\phi}^R_1\ \phi^R_2\})$$

$$\Psi^R_2 = \det\{\phi^R_2\ \bar{\phi}^R_2\} \qquad (13.5.13)$$

$$\Psi^R_3 = \det\{\phi^R_1\ \bar{\phi}^R_1\}$$

The expressions for the matrix elements of $\hat{H}^R_{eff}$ can now be expressed in terms of the AO integrals and the parameters X_R, Y_R for each electron pair. If

$$H^R_{eff,ij} = \int dx_1 \int dx_2\ \Psi^R_i(x_1,x_2)\hat{H}^R_{eff}\ \Psi^R_j(x_1,x_2)$$

then

$$H^R_{eff,11} = I_{11} + I_{22} + (\phi^R_2\phi^R_2,\phi^R_1\phi^R_1)$$

$$H^R_{eff,22} = 2I_{22} + (\phi^R_2\phi^R_2,\phi^R_2\phi^R_2)$$

$$H^R_{eff,33} = 2I_{11} + (\phi^R_1\phi^R_1,\phi^R_1\phi^R_1) \qquad (13.5.14)$$

$$H^R_{eff,12} = \sqrt{2}[I_{12} + (\phi^R_2\phi^R_2,\phi^R_2\phi^R_1)]$$

$$H^R_{eff,23} = (\phi^R_2\phi^R_1,\phi^R_2\phi^R_1)$$

$$H^R_{eff,13} = \sqrt{2}[I_{12} + (\phi^R_2\phi^R_1,\phi^R_1\phi^R_1)]$$

The I_{ij} terms include the one-electron terms and the inter-group repulsion effects.

$$I_{ij} = H_{ij} + \sum_{R\neq S=1}^{n/2} I^{RS}_{ij} + X_S I^{RS}_{x\,ij} + Y_S I^{RS}_{y\,ij} \qquad (13.5.15)$$

where

$$H_{ij} = \int dr\ \phi_i^R(r)\ \hat{h}\ \phi_j^R(r)$$

and

$$I_{ij}^{RS} = [(\phi_i^R\phi_j^R,\phi_1^S\phi_1^S) + (\phi_i^R\phi_j^R,\phi_2^S\phi_2^S) - \tfrac{1}{2}(\phi_i^R\phi_1^S,\phi_j^R\phi_1^S) - \tfrac{1}{2}(\phi_i^R\phi_2^S,\phi_j^R\phi_2^S)]$$

$$I_{x\ ij}^{RS} = [2(\phi_i^R\phi_j^R,\phi_1^S\phi_2^S) - \tfrac{1}{2}(\phi_i^R\phi_1^S,\phi_j^R\phi_2^S) - \tfrac{1}{2}(\phi_i^R\phi_2^S,\phi_j^R\phi_1^S)]$$

$$I_{y\ ij}^{RS} = [(\phi_i^R\phi_j^R,\phi_1^S\phi_1^S) - (\phi_i^R\phi_j^R,\phi_2^S\phi_2^S) - \tfrac{1}{2}(\phi_i^R\phi_1^S,\phi_j^R\phi_1^S) + \tfrac{1}{2}(\phi_i^R\phi_2^S,\phi_j^R\phi_2^S)]$$

The computer implementation of the iterative scheme to calculate the optimum electron-pair molecular wave function is sketched below.

1. Compute the AO molecular integrals and transform them to an orthogonal hybrid basis defined by A of (13.4.2).
2. Estimate initial X_R and Y_R for each electron pair ($X_R = 1$, $Y_R = 0$ corresponds to a non-polar simple MO function for each group).
3. For the R'th group, form the VB matrix whose elements are given by (13.5.14) using the hybrid orbital integrals and the X_R and Y_R parameters for each group.
4. Diagonalise the matrix of H_{eff}^R and obtain a, b and c (the mixing coefficients) of (13.5.11). Hence form new P^R and X_R, Y_R using (13.5.12).
5. Go on to the next electron pair.
6. Are all the X_R, Y_R self-consistent? If not go to (3) otherwise finish.

The principal difficulty with the use of this paired-electron model - as with all VB-type methods - is the necessity of working with an orthogonal basis: the integral transformation bottleneck discussed in Chapter 9. However, inspection of the expressions (13.5.14) and (13.5.15) shows that by no means all of the molecular integrals involving the orthogonal basis are required. The 2×2 block form of the charge and bond-order matrix means that the only off-diagonal H_{ij} elements needed are the ones between orbitals in the same group. Of the m$ electron repulsion integrals, only the intra-group ones (6 distinct integrals per group) and the inter-group ones of the form

$$(\phi_i^R \phi_j^S, \phi_k^R \phi_\ell^S)$$

and $\qquad\qquad\qquad (i,j,k,\ell = 1,2)$

$$(\phi_i^R \phi_j^R, \phi_k^S \phi_\ell^S)$$

are required (10 distinct integrals per pair of groups). By using a partial transformation in which only these necessary integrals are calculated, the separate electron-pair calculation becomes quite feasible.

Using the orthogonal hybrid basis defined by (13.4.6), the solution of the electron-pair problem has been carried through for BeH_2 and the results are presented below. The electron pairs were the Be 1s "lone pair" and the two equivalent Be-H bond pairs (in the case of the 1s shell, of course, only the one function Ψ_3^R was used).

$$P^{1s} = 2$$

$$P^{BeH} = \begin{pmatrix} 1.096 & 0.982 \\ 0.982 & 0.904 \end{pmatrix}$$

Electronic energy = -18.4722 a.u.

There are two main points to note here:

i) the 2×2 charge and bond-order matrix for the bond pair is very similar to the 2×2 block for the transformed LCAOMO calculation (13.4.7) (doubled, since $P_{ij} = 2R_{ij}$).

ii) there is a small absolute gain in energy over the LCAOMO result (0.0091 a.u.) which corresponds to the improvement in the molecular wave function obtained by using a multi-determinant trial function.

It is easy to see that this method should give a better picture of the electron distribution than the LCAOMO method since the LCAOMO method can be thought of as a special case of the electron-pair model in which the "hybrid" orbitals are the MO's and only Ψ_1^R is used for each pair; that is, $\mathsf{A} = \mathsf{U}$ (the linear transformation defining the MO's in terms of the AO's).

13.6 DIRECT CALCULATION OF OPTIMUM "HYBRIDS"

In Section 13.3 we used the results of 13.2 to suggest a method of calculating LMO's without the intervening step of solving the LCAO RHF equation. A completely parallel case can be developed for the calculation of optimum "hybrid" functions. The energy of an electronic system, in the LCAOMO approximation, is given by

$$E = 2\sum_{i,j=1}^{m} H_{ij}R_{ji} + \sum_{i,j,k,\ell=1}^{m}(ij,k\ell)[2R_{ij}R_{k\ell} - R_{ik}R_{j\ell}] \tag{13.6.1}$$

If R is restricted to have 2×2 block form

$$\mathsf{R} = \begin{pmatrix} \mathsf{R}^A & 0 & 0 & \cdots & 0 \\ 0 & \mathsf{R}^B & \cdots & & \cdot \\ \cdot & \cdots & & & \cdot \\ \cdot & \cdots & & & \cdot \\ 0 & & & & \mathsf{R}^R \end{pmatrix} \tag{13.6.2}$$

then, since there is only one degree of freedom for a 2×2 MO problem (which by analogy with 10.3 we choose as $c = \cos\theta$

$$\mathsf{R}^A = \begin{pmatrix} c^2 & 2sc \\ 2sc & s^2 \end{pmatrix} \tag{13.6.3}$$

or,

$$\mathsf{R}^A = \begin{pmatrix} a & 2\sqrt{a(1-a)} \\ 2\sqrt{a(1-a)} & 1-a \end{pmatrix} \tag{13.6.4}$$

Thus (13.6.1) combined with (13.6.2) and (13.6.4) gives an expression containing n/2 parameters for n/2 electron pairs. If each hybrid basis orbitals are approximated by a spherical GTF of variable size and position then the whole expression contains the GTF parameters and the 'a' parameters of (13.6.4). Minimisation of this function gives approximate hybrid functions directly; and localised MO's formed from them are

$$\mathbb{U}^A = \begin{pmatrix} c & -s \\ s & c \end{pmatrix} \qquad (13.6.5)$$

since (13.6.3) holds.

SUGGESTIONS FOR FURTHER READING

"Localised MO's: A Bridge between Chemical Intuition and Molecular Quantum Mechanics" by W. England, L.S. Salmon and K. Ruedenberg is a current review article and appears in "Topics in Current Chemistry" (Springer-Verlag 1971).

The ellipsoidal GTF's described in 13.3 have been used by P.Th. van Duijnen and D.B. Cook for molecular calculations; the work is reported in "Ab initio Calculations with Small Ellipsoidal Gaussian Basis I & II" in Mol.Phys., 21, 475 (1971); 22, 637 (1971).

The use of "floating spherical Gaussian orbitals" (FSGO) for localised descriptions of electronic structure is usually associated with the name of A.A. Frost (e.g. J.Chem.Phys., 47, 3707 (1967)): a recent review article is "Ab initio Calculations on Large Molecules" by R.E. Christoffersen in Advances in Quantum Chem., 6, 333 (1972).

Much semi-qualitative, overlap-based, work has been done on optimum hybrids: an approach suitable for use in quantitative work has been given by R. McWeeny & G. del Re in Theoret.chim. Acta, 10, 13 (1968) "Criteria for Bond Orbitals and Optimum Hybrids".

The theory of separate two-electron groups is taken from "Theory of Separated Electron Pairs" by J.M. Parks and R.G. Parr in J.Chem.Phys., 28, 335 (1958) and also "Self-Consistent

Group Calculations on Polyatomic Molecules" by M. Klessinger & R. McWeeny in J.Chem.Phys., 42, 3343 (1965). The latter reference contains extensions to excited states of electron-pairs and inclusion of "de-localised" groups of electrons.

POST SCRIPT

In trying to bridge the computational gap between quantum theory and chemistry, I have purposely avoided going very deeply into either subject: no specific chemical applications have been discussed and few of the formal properties of the solutions of the Schrödinger equation have been given. I am sure that this is a correct policy for a book (which has become surprisingly large) covering this simple interface area (to use computational jargon). However, molecular wave functions are not computed for their own sake: chemists want chemistry from them and quantum theorists want quantum theory. I have tried to make up for these deficiencies by quoting a *few* references at the end of each chapter in the hope that the interested reader can find his way into the literature.

Some topics have either not been mentioned or dismissed in passing - principally non-MO calculations - and some of these are becoming increasingly important. There has been renewed interest in recent years in VB-like methods but the computation of molecular wave functions outside the orbital model remains a pursuit for only the most determined. These methods will not become routine for quite some time. The principal area not treated in which there is a large amount of current computational interest is in the use of various restricted MCSCF methods. The general formulation of the problem is known and various authors have derived and implemented restricted schemes since the general method is far too expensive in resources for application to molecules of chemical interest. The simultaneous optimisation of orbitals and configuration-mixing coefficients has the work load of both MO and VB calculations. There is a step equivalent to the transformation of electron repulsion integrals to a new orbital basis during every iteration. Many of the restricted MCSCF methods involve giving the "boring" electrons in a molecule (inner shells - but photo-electron spectroscopy is making these electrons distinctly un-boring) an MO wave function and throwing the multi-configurational effort into the valence distribution. This is clearly the next logical step in molecular computations.

On the more "chemical" side of things, I have not discussed approximations *within* the MO model - differential overlap theories. There is scope here for another volume. This type of theory is aimed chiefly at the reliable reproduction of experimental data rather than theoretical interpretations. The continued parallel development of semi-empirical theories and ab initio methods emphasises the complementary nature of experiment and theory in Quantum Chemistry where facts are just as obstinate as in other disciplines.

SUGGESTIONS FOR FURTHER READING

The general MCSCF equation is discussed in "Methods of Molecular Quantum Mechanics" by R. McWeeny & B.T. Sutcliffe (Academic Press 1969) in Section 5.6

Other approachs to the MCSCF problem are "Complete MCSCF Theory" by A. Veillard & E. Clementi in Theoret.chim.Acta, 7, 133 (1967) and "Method of Optimum Valence Configurations" by A.C. Wahl & G. Das in Advances in Quantum Chem., 5, 261 (1970). "Statistical Exchange-Correlation in the Self-Consistent Field" by J.C. Slater in Advances in Quantum Chem., 6, 1 (1972) contains a discussion of methods used in Solid-State theory applied to molecules.

The "cluster" expansion of the electronic wave function is

a useful entry into more generalised treatments of molecular electronic structure; a useful introduction is given in "Separability in Many-Electron Systems" by H. Primas in "Modern Quantum Chemistry - Istanbul Lectures Part II" (Academic Press 1965) Ed. O. Sinanoglu.

Finally, a useful "Bibliography of ab initio Molecular Wave Functions" by W.G. Richards , T.E.H. Walker & R.K. Hinkley is available (Oxford University Press 1971).

APPENDIX A GTF MOLECULAR INTEGRALS

In section 6.6 we used the general Cartesian-orientated form of the GTF's:

$$\chi_i(\alpha_i, \ell_i, m_i, n_i, r_A) = N_i\, x_A^{\ell_i}\, y_A^{m_i}\, z_A^{n_i} \exp(-\alpha_i |r - R_A|^2)$$

where

$$N_i = \left| \frac{2^{2(\ell_i+m_i+n_i)+3/2}\, \alpha^{\ell_i+m_i+n_i+3/2}}{(2\ell_i-1)!!(2m_i-1)!!(2n_i-1)!!} \right|^{\frac{1}{2}}$$

and in Chapter 8 the transformations necessary to evaluate the various molecular integrals were sketched. For the purpose of brevity, in this Appendix the 'χ' is dropped and we treat the GTF as defined by α_i, ℓ_i, m_i, n_i and A the point of origin. Thus, we can write the kinetic energy integral - in an obvious notation - as

$$(\alpha_1,\ \ell_1,\ m_1\ n_1\ A;\ \ -\tfrac{1}{2}\nabla^2;\ \ \alpha_2,\ \ell_2,\ m_2,\ n_2\ B)$$

and use similar shorthand notation for the other molecular integrals. Following 8.4 we choose

$$P = (\alpha_1 r_A + \alpha_2 r_B)/(\alpha_1 + \alpha_2)$$

and

$$Q = (\alpha_3 r_C + \alpha_4 r_D)/(\alpha_3 + \alpha_4)$$

in the two-electron case. In keeping with S.F. Boys' original notation we write the components and magnitudes of the various relative position vectors in the form $\overline{AB}_x$ and $\overline{AB}^2$ etc., where

$$\overline{AB}_x = (r_A - r_B).e_1$$

(e_1 is the unit vector in the x direction) and

$$\overline{AB}^2 = |r_A - r_B|^2$$

This symbolism gives the most transparent interpretation of the formulae in terms of the functions $F_\nu(\tau)$ of 8.4 and the auxiliary functions

$$f_j(\ell,m,a,b) = \sum_{i=\max(0,j-m)}^{\min(j,\ell)} \binom{\ell}{i}\binom{m}{j-1} a^{\ell-i}\, b^{m+i-j}$$

defining the coefficients of powers of x in the expansion of $(x-a)^\ell(x-b)^m$.

As we noted in Chapter 8 the various molecular integrals involving GTF's split into separate integrals over x, y and z. To emphasize this property, symbolic summation signs

$$\sum_{(x)} \qquad \sum_{(y)} \qquad \sum_{(z)}$$

have been used which indicate how the integral has been broken down. Specification of one of these symbols in each case is enough since the expressions for x, y and z summations are precisely analogous. The symbol $\lfloor i$ (the "floor" of i) is used to denote the largest integer $\leq i$.

THE OVERLAP INTEGRAL

$$(\alpha_1,\ell_1,m_1,n_1,A;\ \alpha_2,\ell_2,m_2,n_2,B)$$

$$= N_1N_2\left(\frac{\pi}{\alpha_1+\alpha_2}\right)^{3/2} \exp\left(-\frac{\alpha_1\alpha_2}{\alpha_1+\alpha_2}\overline{AB}^2\right)$$

$$\times \sum_{(x)} G_i^x \sum_{(y)} G_j^y \sum_{(z)} G_k^z$$

where

$$\sum_{(x)} = \sum_{i=0}^{\lfloor(\ell_1+\ell_2)/2}$$

and

$$G_i^x = \mathbf{f}_{2i}(\ell_1, \ell_2, \overline{PA}_x, \overline{PB}_x) \frac{(2i-1)!!}{2^i(\alpha_1+\alpha_2)^i}$$

etc.

THE KINETIC ENERGY INTEGRAL

$$(\alpha_1,\ell_1,m_1,n_1,A;\ -\tfrac{1}{2}\nabla^2;\ \alpha_2,\ell_2,m_2,n_2,B)$$

$$= N_1N_2[\alpha_2\{2(\ell_2+m_2+n_2)+3\}(\alpha_1,\ell_1,m_1,n_1,A;\alpha_2,\ell_2,m_2,n_2,B)$$

$$- 2\alpha_2^2\{(\alpha_1,\ell_1,m_1,n_1,A;\alpha_2,\ell_2+2,m_2,n_2,B)$$

$$+ (\alpha_1,\ell_1,m_1,n_1,A;\alpha_2,\ell_2,m_2+2,n_2,B)$$

$$+ (\alpha_1,\ell_1,m_1,n_1,A;\alpha_2,\ell_2,m_2,n_2+2mB)\}$$

$$- \tfrac{1}{2}\ \ell_2(\ell_2-1)\times(\alpha_1,\ell_1,m_1,n_1,A;\alpha_2,\ell_2-2,m_2,n_2,B)$$

$$- \tfrac{1}{2}\ m_2(m_2-1)\times(\alpha_1,\ell_1,m_1,n_1,A;\alpha_2,\ell_2,m_2-2,n_2,B)$$

$$- \tfrac{1}{2}\ n_2(n_2-1)\times(\alpha_1,\ell_1,m_1,n_1,A;\alpha_2,\ell_2,m_2,n_2-2,B)]$$

This is the general expansion for the kinetic energy integral in terms of the seven possible overlap integrals generated by the differentiations. The expression looks unsymmetrical in that ℓ_1, m_1, n_1 appear in a form different from the appearances of ℓ_2, m_2, n_2: in fact the integral shows correct

Hermitian symmetry. If any (or all) of ℓ_2, m_2, n_2 are less than 2 then the relevant selection of the last three terms is omitted.

THE NUCLEAR ATTRACTION INTEGRAL

$$(\alpha_1, \ell_1, m_1, n_1, A; \ \frac{1}{r_C}\ ; \ \alpha_2, \ell_2, m_2, n_2, B)$$

$$= N_1 N_2 \left(\frac{\pi}{\alpha_1+\alpha_2}\right) \exp\left(-\frac{\alpha_1\alpha_2}{\alpha_1+\alpha_2}\overline{AB}^2\right)$$

$$\times \sum_{(x)} A^x_{i,r,u} \sum_{(y)} A^y_{j,s,v} \sum_{(z)} A^z_{k,t,w} F_\nu(\tau)$$

where

$$\sum_{(x)} = \sum_{i=0}^{\ell_1+\ell_2} \sum_{r=0}^{\lfloor i/2} \sum_{u=0}^{\lfloor (i-2r)/2} \qquad \text{etc.}$$

and

$$A^x_{i,r,u} = (-1)^{i+u} f_i(\ell_1, \ell_2, \overline{PA}_x, \overline{PB}_x) \frac{i!\,\overline{CP}_x^{\,i-2(r+u)}\varepsilon^{r+u}}{r!u!(i-2r-2u)!}$$

$$\varepsilon = 1/4(\alpha_1 + \alpha_2); \quad \tau = (\alpha_1 + \alpha_2)\overline{CP}^2$$

$$\nu + i + j + k - 2(r+s+t) - u - v - w \qquad \text{etc.}$$

THE ELECTRON REPULSION INTEGRAL

$$(\alpha_1, \ell_1, m_1, n_1, A; \ \alpha_2, \ell_2, m_2, n_2, B; \ \frac{1}{r_{12}}; \ \alpha_3, \ell_3, m_3, n_3, C;$$

$$\alpha_4, \ell_4, m_4, n_4, D)$$

$$= N_1 N_2 N_3 N_4 \frac{2\pi^2}{\gamma_1\gamma_2}\left(\frac{\pi}{\gamma_1+\gamma_2}\right) \exp\left(-\frac{\alpha_1\alpha_2\overline{AB}^2}{\gamma_1} - \frac{\alpha_3\alpha_4\overline{CD}^2}{\gamma_2}\right)$$

$$\sum_{(x)} B^x_{i_1,i_2,r_1,r_2,u} \quad \sum_{(y)} B^y_{j_1,j_2,s_1,s_2,v}$$

$$\sum_{(z)} B^z_{k_1,k_2,t_1,t_2,w} \quad F_\nu(\tau)$$

where

$$\sum_{(x)} = \sum_{i_1=0}^{\ell_1+\ell_2} \sum_{i_2=0}^{\ell_3+\ell_4} \sum_{r_1=0}^{\llcorner i_1/2} \sum_{r_2=0}^{\llcorner i_2/2} \sum_{u=0}^{u'} \qquad \text{etc.}$$

$$u' = \llcorner((i_1+i_2)/2 - r_1 - r_2)$$

$$\gamma_1 = \alpha_1 + \alpha_2 \; ; \quad \gamma_2 = \alpha_3 + \alpha_4 \; ;$$

$$\tau = \overline{PQ}^2/4(\gamma_2^{-1} + \gamma_2^{-1})$$

$$\nu = i_1 + i_2 + j_1 + j_2 + k_1 + k_2 - 2(r_1+r_2+s_1+s_2+t_1+t_2)-u-v-w$$

and

$$B^x_{i_1,i_2,r_1,r_2,u} = (-1)^{i_2+u} \; f_{i_1}(\ell_1,\ell_2,\overline{PA}_x,\overline{PB}_x) f_{i_2}(\ell_3,\ell_4.\overline{QC}_x,\overline{QD}_x)$$

$$\times \frac{i_1!i_2!\,\gamma_1^{r_1-i_1}\,\gamma_2^{r_2-i_2}\,(2\delta)^{2(r_1+r_2)}}{r_1!r_2!(i_1-2r_1)!(i_2-2r_2)!(4\delta)^{i_1+i_2}}$$

$$\times \frac{(\llcorner(i_1+i_2-2r_1-2r_2))!\;\overline{PQ}_x^{\,i_1+i_2-2(r_1+r_2+u)}\;\delta^u}{u!(\llcorner(i_1+i_2-2(r_1+r_2+u)))!}$$

$$\delta = \frac{1}{4}\left(\frac{1}{\gamma_1} + \frac{1}{\gamma_2}\right) \qquad \text{etc.}$$

APPENDIX B PROGRAM PACKAGES

The following is a short list of large-scale molecular wave function program "packages" which are available and are operational in many computing laboratories throughout the world. Typically, these programs are written for efficiency, not clarity, (contrasting perhaps with the routines listed in Appendix C) and it is not easy for the beginner to amend them or to adapt them to his own research needs. The programs, as written, demand very powerful computing facilities and cannot be run in un-modified form on the average (British) University Computer.

1. Atomic RHF Calculations: IBM Technical Report RJ518 by B. Roos, C. Salez, A. Veillard and E. Clementi. (Available on request from IBM Research Laboratory, San Jose, California 95114, U.S.A.). This program uses STO's or GTF's.
2. Diatomic Molecule Calculations: BISON by A.C. Wahl, P.J. Bertonani, K. Kauser and R.H. Land, Argonne National Laboratory Technical Report ANL 7271 (Available on request from Argonne National Laboratory, 9700 South Cass Avenue, Illinois 60439, U.S.A.). The program uses STO functions.
3. Linear Molecule Calculations: McL-YOSH Linear Molecule Program by A.D. McLean and M. Yoshimine (Also ALCHEMY by the same authors). Available on request from

Quantum Chemistry Program Exchange (QCPE) Indiana University*. STO functions are used by this program package.

4. General Molecular Calculations:
   a) POLYATOM 2 by D.B. Neumann, H. Basch, R.L. Korregay, L.C. Snyder, J. Moskowitz, C. Hornbuck and P. Liebmann. Available on request from QCPE*. This suite of programs uses GTF's.
   b) IBMOL IV: IBM Technical Report "IBMOL Version 4" by A. Veillard. Available on request from IBM, San Jose. GTF basis is used.
   c) ATMOL by I.H. Hillier and V. Saunders is a modified IBMOL - documentation is available from SRC ATLAS Laboratory, Chilton, Didcot, Berks, U.K.

No use of molecular symmetry is made in IBMOL (or ATMOL): an algorithm which makes use of a single plane of symmetry is described by P. Seigbahn in Chem.Phys.Letters, 8, 246 (1971).

---

* QCPE makes a handling charge for distributing programs: full information can be had from QCPE, Chemistry Department, Room 204, Indiana University, Bloomington, Indiana 47401, U.S.A.

APPENDIX C SAMPLE PROGRAMS

The sample programs and program fragments scattered through the text are used to illustrate the ideas developed there and are not particularly efficient and certainly do not form a complete system for molecular calculations. In this Appendix a collection of routines - extensively documented with "comments" statements - are given which are being used in a routine way for closed shell molecular calculations (together with implementations of the formulae in Appendix A). These routines are segmented in a logical way - one subroutine per operation - and are designed to be used as a *basis* into which the computational worker can insert his own ideas. It is hoped that the reader will go through the main routines (at least in outline by following the "comments") and compare the programming to the ideas developed in the text. This is by far the best way of getting a "feel" for how calculations are organised.

There are a few points to be made about the methods used in the whole system.

i) The IBM Scientific Subroutine Package (IBM SSP - IBM Document No. H20-0166) method of storing matrices in a one-dimensional array is used throughout - experience has shown that "singly subscripted FORTRAN" compiles to quite efficient object code. The method is described in SUBROUTINE LOC: briefly, all matrices are stored in column order, symmetric matrices by columns to the main diagonal and general matrices by full columns. The two routines LOC and MODE facilitate matrix handling.

ii) The storage space required for a file of electron repulsion integral labels and values is by an integer "packing" procedure. The routine PACK stores all four labels I, J, K and L in one real FORTRAN variable and UNPAC reverses this procedure. This saving of space is not too important except in the largest calculations and in any case is only around 25%. Packing is a cumbersome procedure in FORTRAN and so PACK and UNPAC are the only lapses from FORTRAN.

iii) The routines are all written in Basic FORTRAN and should compile and run on any system boasting a FORTRAN compiler.

iv) All routines are "free standing" - the only interface is through the explicit arguments, and are (hopefully) self-explanatory. The *assumed* FUNCTIONS OEI and ERI are exceptions; if the integral formulae are programmed then OEI and ERI must communicate data via COMMON (a sensible modification to DHLST and DGLST would be to make OEI and ERI EXTERNAL).

v) The routines have been designed to run in a minimum of fast store (about 8K words - 32K bytes) and therefore rather heavy use has been made of a direct access device for temporary storage of matrices; this may slow the programs down in certain operating environments. However, it is increasingly common for time-sharing operating systems to use "paging". Paging involves parts of the user's program being written to backing store at times which are, from the user's point of view, arbitrary and so the use of a direct access device in a logical way can increase efficiency.

```
      SUBROUTINE DGLST(N,MTRAN,MMAX,NFILE)
C  ....  ROUTINE TO COMPUTE AND STORE THE DISTINCT TWO ELECTRON
C  ....  INTEGRALS WHEN THE MOLECULAR POINT GROUP IS SUMMARISED IN
C  ....  MTRAN - OUTPUT ONLY SUITABLE FOR USE WITH GOFR CALLING SYMG
C  ....  NOTES ..
C  ....   MTRAN MUST NOT CONTAIN THE 'IDENTITY' COLUMN BUT MUST
C  ...  SATISFY THE USUAL 'GROUP' CONDITIONS.
C  ....  MTRAN MUST NOT CONTAIN ANY ZEROES - FOR MOLECULES WITH
C  ....  A PRINCIPLE C3 OR C5 AXIS A SUB-GROUP FOR WHICH MTRAN
C  ....  HAS NO ZEROES  MUST BE USED.
C*********************************************************************C
C   ARGUMENTS                                                         C
C          N   THE NUMBER OF BASIS ORBITALS                           C
C          MTRAN  THE NUMBER MTRAN(I,J) IS THE NUMBER OF THE ORBITAL  C
C                 INTO WHICH ORBITAL  I  IS TRANSFORMED  BY SYMMETRY  C
C                 OPERATION  J  OF THE POINT GROUP - NO ZEROES ALLOWED C
C          MMAX   THE NUMBER OF OPERATIONS IN THE POINT GROUP -       C
C                 EXCLUDING THE IDENTITY                              C
C          NFILE  THE LOGICAL NUMBER OF A FILE FOR OUTPUT OF THE      C
C                 DISTINCT TWO ELECTRON INTEGRALS                     C
C*********************************************************************C
      REAL LABEL(200)
      DIMENSION MTRAN(40,24),VALUE(200)
      DATA NN/200/
      IEND=0
      NEXT=0
      DO 1 I=1,N
      DO 1 J=1,I
      DO 1 K=1,I
      LTOP=K
      IF(I-K) 3,2,3
    2 LTOP=J
    3 DO 1 L=1,LTOP
      IF(L-N) 7,8,8
    8 IEND=1
    7 NUM=1
      PROD=FLOAT(I*J)*FLOAT(K*L)
      P=P4(I,J,K,L)
      DO 4 M=1,MMAX
      IT=MTRAN(I,M)
      JT=MTRAN(J,M)
      KT=MTRAN(K,M)
      LT=MTRAN(L,M)
      TPROD=FLOAT(IT*JT)*FLOAT(KT*LT)
      CALL ORDER(IT,JT,KT,LT)
      PT=P4(IT,JT,KT,LT)
      IF(P-PT) 6,6,1
    6 IF(I-IT) 4,44,4
   44 IF(J-JT) 4,45,4
   45 IF(K-KT) 4,46,4
   46 IF(L-LT) 4,47,4
   47 IF(TPROD) 4,49,48
C  ....  NUM COUNTS THE RECURRANCES OF (IJ,KL)
   48 NUM=NUM+1
    4 CONTINUE
      NEXT=NEXT+1
      CALL PACK(LABEL(NEXT),I,J,K,L)
C  ....  COMPUTE A 'SCALED' DISTINCT (IJ,KL)
      VALUE(NEXT)=ERI(I,J,K,L)/FLOAT(NUM)
      IF(IEND) 10,10,9
   10 IF(NEXT-NN) 1,9,9
C  ....  WRITE OUT A FULL BUFFER
    9 WRITE(NFILE) NEXT,IEND,LABEL,VALUE
      NEXT=0
    1 CONTINUE
C  ....  PUT MTRAN ON THE END OF THE FILE FOR SYMG
      WRITE(NFILE) MMAX,MTRAN
      RETURN
C  ....  '49' IS AN ERROR CONDITION - MTRAN SHOULD HAVE NO ZEROES
   49 STOP
      END
```

```
      SUBROUTINE DHLST(N,MTRAN,MMAX,H)
C  ....  ROUTINE TO COMPUTE AND STORE (IN H) ONLY THE 'POINT GROUP
C  ....  DISTINCT' ONE ELECTRON INTEGRALS - FOR USE WITH GOFR CALLING
C  ....  SYMG (AND NOT OTHERWISE).
C  ....  SEE DGLST FOR COMMENTS ON THE RESTRICTIONS ON MTRAN.
C*************************************************************************C
C       ARGUMENTS                                                          C
C           N  THE NUMBER OF BASIS ORBITALS                                C
C           MTRAN    MTRAN(I,J) IS THE ORBITAL INTO WHICH ORBITAL I        C
C                   IS TRANSFORMED BY SYMMETRY OPERATION J - WITH SIGN     C
C           MMAX    THE NUMBER OF OPERATIONS IN THE POINT GROUP            C
C                      MTRAN MUST NOT CONTAIN THE IDENTITY                 C
C           H     A STORAGE MODE 1 MATRIX FOR OUTPUT                       C
C*************************************************************************C
      DIMENSION H(1),MTRAN(40,1)
      DATA ZERO/0.0/
      NS=N*(N+1)/2
C  ....  INITIALISE H MATRIX
      DO 20 I=1,NS
   20 H(I)=ZERO
      DO 7 I=1,N
      DO 7 J=1,I
      IJ=I*(I-1)/2+J
      IPR=I*J
      NUM=1
      DO 4 M=1,MMAX
      IT=MTRAN(I,M)
      JT=MTRAN(J,M)
      ITPR=IT*JT
      IT=IABS(IT)
      JT=IABS(JT)
      IF(IT-JT) 41,42,42
   41 ID=IT
      IT=JT
      JT=ID
   42 IJT=IT*(IT-1)/2+JT
      IF(IJ-IJT) 6,6,7
    6 IF(I-IT) 4,44,4
   44 IF(J-JT) 4,45,4
   45 IF(ITPR) 7,49,46
C  ....  NUM COUNTS THE NUMBER OF TIMES H(IJ) IS SENT INTO
C  ....        ITSELF
   46 NUM=NUM+1
    4 CONTINUE
C  ....  COMPUTE A 'SCALED' DISTINCT H(IJ) ELEMENT
      H(IJ)=OEI(I,J)/FLOAT(NUM)
    7 CONTINUE
      RETURN
C  ....  '49' IS AN ERROR CONDITION - MTRAN SHOULD HAVE NO ZEROES
   49 STOP
      END


      FUNCTION P4(I,J,K,L)
C  ....  FORMS THE BOOK-KEEPING NUMBER FROM I,J,K,L IN STANDARD ORDER
      IP2F(M,N)=N+(M*(M-1))/2                                            P4
      P2IJ=FLOAT(IP2F(I,J))
      P2KL=FLOAT(IP2F(K,L))
      P4=P2KL+0.5*P2IJ*(P2IJ-1)
      RETURN
      END
```

```
      SUBROUTINE LOC(I,J,IR,N,M,MS)
C   ....  GENERAL ROUTINE  FOR BOOK-KEEPING OF SINGLY SUBSCRIPTED
C   ....  MATRICES - WORKS FOR STORAGE MODE 0(GENERAL),1(SYMMETRIC)
C   ....  AND 2(DIAGONAL).
C   ....  STORAGE IS IN ALL CASES BY COLUMNS - VIZ
C   ....   FOR A GENERAL MATRIX THE (I,J) ELEMENT IS A((J-1)*N+I)
C   ....   FOR A SYMMETRIC MATRIX (I,J) ELEMENT IS A(I*(I-1)/2+J)
C   ....  WHERE I.GE.J  OR VICE VERSA FOR J.GT.I
C   ....  FOR DIAGONAL MATRICES (I,I) ELEMENT IS STORED AT A(I)
C*********************************************************************C
C    ARGUMENTS                                                        C
C           I    ROW POSITION OF ELEMENT                              C
C           J   COLUMN POSITION OF ELEMENT                            C
C           IR    COMPUTED SINGLE SUBSCRIPT                           C
C           N       NO. OF ROWS IN DATA MATRIX                        C
C           M    NO. OF COLUMNS IN DATA MATRIX                        C
C           MS      STORAGE MODE OF DATA MATRIX                       C
C*********************************************************************C
      IX=I
      JX=J
      IF(MS-1) 10,20,30
   10 IRX=N*(JX-1)+IX
      GO TO 36
   20 IF(IX-JX) 22,24,24
   22 IRX=IX+(JX*JX-JX)/2
      GO TO 36
   24 IRX=JX+(IX*IX-IX)/2
      GO TO 36
   30 IRX=0
      IF(IX-JX) 36,32,36
   32 IRX=IX
   36 IR=IRX
      RETURN
      END
```

```
      SUBROUTINE MODE (A,N,IN,IOUT)
      DIMENSION A(1)
C   ....  CHANGES STORAGE MODE      OF A FROM IN TO IOUT (BOTH IN AND       MOD
C   ....  IOUT MUST BE 0 OR 1 )
C   ....  USAGE IS IN MAKING SYMMETRICAL MATRICES USABLE BY VDGHV
C   ....  AND MATP
C*********************************************************************C
C    ARGUMENTS                                                        C
C           A   MATRIX TO BE MANIPULATED                              C
C           N   DIMENSION OF A                                        C
C           IN   STORAGE MODE OF A AS INPUT (0 OR 1)                  C
C           IOUT  STORAGE MODE OF A AS OUTPUT (1 OR 0)                C
C*********************************************************************C
      IF(IN)1,2,1
    2 DO 3 J=1,N
      DO 3 I=1,J
      CALL LOC(I,J,IJ0,N,N,0)
      CALL LOC(I,J,IJ1,N,N,1)
    3 A(IJ1)=A(IJ0)
      RETURN
    1 CONTINUE
      DO 4 JB=1,N
      J=N+1-JB
      DO 4 IB=1,J
      I=J+1-IB
      CALL LOC(I,J,IJ0,N,N,0)
      CALL LOC(I,J,IJ1,N,N,1)
    4 A(IJ0)=A(IJ1)
      DO 5 J=1,N
      DO 5 I=1,J
      CALL LOC(I,J,IJ,N,N,0)
      CALL LOC(J,I,JI,N,N,0)
    5 A(JI)=A(IJ)
      RETURN
      END
```

```
      SUBROUTINE SCFCS(A,B,N,NDOCC,IBLOC,NBLOC,NFILE,IFILE,NSCRA,
     * ISCRA,IRITE,ICARD)
C   ....  CLOSED SHELL SCF - LC' AO'  PROGRAM
C   ....  SEE CHAPTER 9 FOR THEORY.
C   ....  THE  V  MATRIX SUPPLIED IN  B  IS ASSUMED TO TRANSFORM
C   ....  THE HF MATRIX INTO AN ORTHOGONAL BASIS OF SYMMETRY ORBITALS
C   ....  AFTER THIS TRANSFORMATION THE HF MATRIX MUST BE 'BLOCKED'
C   ....  I.E. CONSIST OF  NBLOC SUB-MATRICES OF DIMENSION
C   ....  IBLOC(1) ..... IBLOC(NBLOC)    ALONG THE MAIN DIAGONAL.
C   ....   IF THE V  MATRIX IS MERELY S**-0.5  SET NBLOC 1,IBLOC(1) =N
C   ....    THE V  MATRIX WILL GENERALLY BE  ....
C   ....       V= T*(TDAG*S*T)**0.5     WHERE T PRODUCES SYMMETRY ORBITALS
C   ....  THE WORKING MATRICES ARE GIVEN NEUTRAL NAMES TO AVOID
C   ....  NOMENCLATURE CONFUSION E.G.  WHEN  'H' CONTAINS  R    ETC.
      DIMENSION A(640),B(640),IBLOC(1)
C   ....  A SENSIBLE MODIFICATION WOULD BE TO MAKE THE MATRICES A,B OF
C   ....  'VARIABLE' DIMENSION -ADDING AN ARGUMENT TO THE LIST
C   ....       IN THIS  CASE N*N
C*********************************************************************C
C     ARGUMENTS                                                       C
C            A    THE ONE ELECTRON HAMILTONIAN - STORAGE MODE 1       C
C           B   THE V MATRIX WHICH ORTHOGONALISES AND SYMMETRY        C
C                 BLOCKS THE HF  MATRIX - STORAGE MODE 0 OF COURSE    C
C           N     DIMENSION OF A,B                                    C
C            NDOCC      NUMBER OF DOUBLY OCCUPIED ORBITALS            C
C           IBLOC     VECTOR CONTAINING THE SUB MATRIX INFORMATION    C
C           NBLOC THE NUMBER OF SUB-MATRICES IN THE TRANSFORMED HF    C
C                              MATRIX                                 C
C            NFILE    THE  LOGICAL NUMBER OF THE TWO ELECTRON FILE    C
C           IFILE    THE ASSOCIATED VARIABLE FOR NFILE                C
C                      IFILE IS REDUNDANT IN NON-RANDOM ACCESS FILES  C
C           NSCRA    THE LOGICAL NUMBER OF A DISK SCRATCH FILE  MUST  C
C                     HAVE ENOUGH ROOM FOR SIX  N*N MATRICES          C
C           ISCRA      THE ASSOCIATED VARIABLE FOR NSCRA              C
C           IRITE   THE LOGICAL NUMBER OF A PRINTER ( 3 ON 1130)      C
C           ICARD    THE LOGICAL NUMBER OF A CARD PUNCH OR ZERO IF    C
C                    NO CARD OUTPUT OF R,U  IS WANTED  (2 OR 0 ON 1130 C
C*********************************************************************C
      DIMENSION COL(50)
      DATA R,U/'R','U' /
C   ....  CONSTANTS ARE PUT INTO 'DATA' STATEMENTS TO FACILITATE
C   ....  CHANGEOVER TO DOUBLE PRECISION - IMPLICIT REAL*8
      DATA CRIT,ZERO,ONE,TWO/1.0E-04,0.0,1.0,2.0/
C
C   ....  REMOVE THE FOLLOWING ARITHMETIC STATEMENT FUNCTION WHEN
C   ....  CHANGING TO DOUBLE PRECISION
      DABS(X)=ABS(X)
      NN=N*N
      NS=N*(N+1)/2
      ISCRA=1
      WRITE(NSCRA'ISCRA)  B
C   ....  FOR MTS USAGE TAKE THE 'C' FROM THE FOLLOWING CARD
C     ISCRA=2
C   ....  WORK  OUT STORAGE POSITIONS FOR THE VARIOUS MATRICES
      IVPOS=1
      IHPOS=ISCRA
      IRPOS=2*ISCRA-1
      IUPOS=3*ISCRA-2
      IEPOS=4*ISCRA-3
      WRITE(NSCRA'IHPOS)  A
C
      FIND (NSCRA'IRPOS)
C   ....  COMPUTE AND STORE A ZERO INITIAL R  MATRIX
C   ....  FEEL FREE TO MAKE ANY OLD GUESS AT 'R' TO START
      DO 1 I=1,NN
    1 B(I)=ZERO
      WRITE(NSCRA'IRPOS) B
C
      JUMP=1




C   ....  THIS 'BOX'  CONTAINS THE ITERATIVE PROCEDURE
```

```
C''''''''''''''''''''''''''''''''''''''''''''''''''''''''''''''''''''''''''''''
   10 READ(NSCRA'IHPOS) A
      ICON=0
      EN=ZERO
C   ....  H  MATRIX  IN  A
C   ....  THE ENERGY CALCULATION IS MEANINGLESS ON THE FIRST CYCLE
      GO TO (8,9), JUMP
    9 CONTINUE
C   ....  THIS CALL TO SYMG IS TO GET THE ENERGY RIGHT
      CALL SYMG(A,B,N,NFILE)
      READ(NSCRA'IRPOS) B
C   ....  R  MATRIX  IN B
C   ....  THE ELECTRONIC ENERGY IS ACCUMULATED IN EN
      CALL TRACE (A,B,N,EN)
      READ(NSCRA'IHPOS) A
    8 JUMP=2
      FIND(NSCRA'IVPOS)
C   ....   GOFR  SCANS THE FILE  NFILE  AND FORMS G(R)
      CALL GOFR(B,A,N,NFILE,IFILE,TWO,ONE)
      READ(NSCRA'IRPOS) B
      CALL TRACE(A,B,N,EN)
C   ....        THE TOTAL ELECTRONIC ENERGY IS GIVEN BY  .....
C   ....        E =      (TRACE R*H + TRACE R*HF )
C   ....  THERE IS, OF COURSE, NO NEED TO COMPUTE EN EVERY CYCLE
C   ....  ONCE AT THE END IS ENOUGH
C   ....  NON-ORTHOGONAL  HF  NOW IN A
      READ(NSCRA'IVPOS) B
C   ....  ORTHOGONALISING  MATRIX  NOW IN B
      CALL MODE (A,N,1,0)
C   ....  EXPAND OUT HF  TO FULL STORAGE MODE FOR  VDGHV
C   ....  VDGHV  TRANSFORMS THE HF MATRIX TO ORTHOGONAL SYMMETRY ORBITALS
      CALL VDGHV(A,B,COL,N)
      CALL MODE (A,N,0,1)
C   ....  ORTHOGONAL  HF  NOW  IN A  AND CHANGED TO  MODE 1
      FIND (NSCRA'IUPOS)
C   ....  DIAGONALISE THE ORTHOGONAL HF MATRIX
      CALL EIGEB(A,B,N,0,IBLOC,NBLOC)
      WRITE(NSCRA'IUPOS) B
C   ....  THE ORTHOGONAL EIGENVECTORS WRITTEN TO FILE
      WRITE(NSCRA'IEPOS) A
C    ....  THE EIGENVALUES WRITTEN TO FILE AS THE DIAGONALS OF  A
      READ(NSCRA'IVPOS) A
C   ....  PICK UP THE ORTHOGONALISING MATRIX TO FORM THE AO EIGENVECTORS
C   ....  THE 'AO' EIGENVECTORS ARE RELATED TO THE ORTHOGONAL ONES
C   ....   BY -  U = V*UBAR    (UBAR ORTHOGONAL VECTORS)
      FIND (NSCRA'IRPOS)
      CALL MM(A,B,COL,N)
C   ....  B  NOW CONTAINS THE NON-ORTHOGONAL EIGENVECTORS
C   ....  RMAT FORMS THE DENSITY MATRIX FROM THE OCCUPIED COLUMNS OF U
      CALL RMAT(B,A,N,1,NDOCC)
C   ....  A  NOW CONTAINS THE NON-ORTHOGONAL  R  MATRIX
C
      READ(NSCRA'IRPOS) B
C   ....  PICK UP THE PREVIOUS  R  MATRIX
C   ....  AND  TEST FOR CONVERGENCE
C   ....  THIS IS THE POINT TO INSERT YOUR PET EXTRAPOLATION METHOD
      SUM=ZERO
      DO 2 I=1,NS
      TERM=DABS(A(I)-B(I))
      IF(TERM-CRIT) 2,3,3
    3 ICON =1
    2 SUM=SUM+TERM
C   ....  WRITE  BACK THE NEW  R  MATRIX
      WRITE(NSCRA'IRPOS) A
C
C   ....  ICON REMAINING ZERO INDICATES CONVERGENCE
      WRITE(IRITE,100) SUM
  100 FORMAT ('   SUM OF DIFFERENCES  IN  R =  ',E20.8)
      WRITE(IRITE,102) EN
      IF(ICON) 20,20,10
   20 CONTINUE
C   ....  THIS  IS THE END OF THE ITERATIVE PROCEDURE - THE REST OF
C   ....  THE PROGRAM  IS  OUTPUT OF THE RESULTS
C''''''''''''''''''''''''''''''''''''''''''''''''''''''''''''''''''''''''''''''
C
      FIND (NSCRA'IUPOS)
```

```
      CALL MODE (A,N,1,0)
C  ....  OUTPUT  THE SELF CONSISTENT NON-ORTHOGONAL  R  MATRIX
      CALL MATP(A,N,N,N,R,IRITE)
      READ(NSCRA'IUPOS) A
      FIND (NSCRA'IVPOS)
      READ(NSCRA'IVPOS) B
      FIND(NSCRA'IEPOS)
      CALL MM(B,A,COL,N)
C  ....  OUTPUT THE AO EIGENVECTORS
      CALL MATP(A,N,N,N,U,IRITE)
      READ(NSCRA'IEPOS) A
      FIND (NSCRA'IRPOS)
C  ....  SORT OUT THE DIAGONALS OF  HF
      DO 4 I=1,N
      II=I*(I+1)/2
    4 B(I)=A(II)
      WRITE(IRITE,101) (B(I),I=1,N)
  101 FORMAT ('  ORBITAL ENERGIES   .....',/,(5F20.7))
      WRITE(IRITE,102) EN
  102 FORMAT('    TOTAL ELECTRONIC ENERGY =  ',E20.8)
      IF(ICARD) 22,22,21
   21 READ(ICARD,200)
      READ(NSCRA'IRPOS) A
      FIND (NSCRA'IUPOS)
      WRITE(ICARD,200) (A(I),I=1,NS)
      READ(NSCRA'IUPOS) A
      WRITE(ICARD,200) (A(I),I=1,NN)
  200 FORMAT (8F10.6)
   22 CONTINUE
C  ....   RESTORE THE ARGUMENTS  A, B
      READ(NSCRA'IHPOS) A
      READ(NSCRA'IVPOS) B
      RETURN
C     $$$$$$$$$$$$$$$$$$$$$$$$$$$$$$$$$$$$$$$$$$$$$$$$$$$$$$$$$$
C     $      CALLED  SUBROUTINES                                 $
C     $     GOFR, TRACE, MODE , VDGHV, RMAT,                     $
C     $     EIGEB, MATP, UNPAC, LOC ,MM                          $
C     $$$$$$$$$$$$$$$$$$$$$$$$$$$$$$$$$$$$$$$$$$$$$$$$$$$$$$$$$$
```

```
      SUBROUTINE GOFR(R,G,N,NFILE,IFILE,EM,EN)
C  ....  ADDS G(R) = EM*J(R) - EN*K(R) INTO  G
C  ....  I.E. THE ONE ELECTRON HAMILTONIAN WILL NORMALLY BE IN  G
C  ....  ON CALLING THIS ROUTINE
C*********************************************************************C
C    ARGUMENTS                                                        C
C          R   INPUT DENSITY MATRIX - STORAGE MODE 1                  C
C          G   G(R) IS ADDED TO THE MATRIX  G                         C
C          N   THE  DIMENSION  OF  R,G                                C
C          NFILE  THE LOGICAL NUMBER OF THE TWO ELECTRON INTEGRAL     C
C                      FILE                                           C
C          IFILE   REDUNDANT - CAN BE USED AS THE ASSOCIATED FILE     C
C                 VARIABLE  IN DISK APPLICATIONS                      C
C          EM,EN  THE MULTIPLIERS OF THE COULOMB AND EXCHANGE         C
C                   MATRICES IN THE DEFINITION OF G(R)  AS ABOVE      C
C*********************************************************************C
      REAL LABEL(200)
      DIMENSION VALUE(200)
      DIMENSION R(1),G(1)
      REWIND NFILE
      IFILE =1
C  ....  FOR DISK APPLICATIONS TAKE THE 'C' FROM THE FOLLOWING CARD
C  51 READ(NFILE'IFILE) NN,IEND,LABEL,VALUE
C  ....  AND THROW AWAY THE PRESENT STATEMENT 51
   51 READ(NFILE) NN,IEND,LABEL,VALUE
      DO 29 M=1,NN
      CALL UNPAC(LABEL(M),I,J,K,L)
C  ....  'REVERSE DICTIONARY' ORDER
C  ....  COMPUTE THE SUBSCRIPTS FOR THE CONTRIBUTIONS OF I,J,K,L
C  ....  TO THE G MATRIX
      II=I*(I-1)/2
      KK=K*(K-1)/2
      IJ=II+J
      KL=KK+L
      CALL LOC(J,K,JK,N,N,1)
      IL=II+L
      IK=II+K
      CALL LOC(J,L,JL,N,N,1)
      VA=VALUE(M)
      VAL =VA*EM
      RN=VA*EN
      DA=R(IJ)*VAL
      DB=R(KL)*VAL
      SV=R(JK)*RN
      SA=R(IK)*RN
      SB=R(IL)*RN
      SC=R(JL)*RN
      IF(K-L) 99,97,99
   99 DB=DB+DB
      G(IK)=G(IK)-SC
      IF(I-J) 96,97,96
   96 IF(J-K) 97,95,95
   95 G(JK)=G(JK) -SB
   97 G(IL)=G(IL)-SV
      G(IJ)=G(IJ)+DB
      IF(I-J) 14,15,14
   14 IF(J-L) 15,12,12
   12 G(JL)=G(JL)-SA
   15 IF(IJ-KL) 21,29,21
   21 CON=DA
      IF(I-J) 56,57,56
   56 CON=CON+DA
   57 IF(J-K) 61,61,22
   61 G(JK)=G(JK)-SB
      IF(I-J) 59,62,59
   59 IF(I-K) 60,60,62
   60 G(IK)=G(IK)-SC
   62 IF(K-L) 74,22,74
   74 IF(J-L) 73,73,22
   73 G(JL)=G(JL)-SA
   22 G(KL)=G(KL)+CON
   29 CONTINUE
      IF(IEND) 53,51,53
   53 CONTINUE
C
```

```
C  ....  TO USE A FILE OF POINT GROUP DISTINCT INTEGRALS ONLY -
      CALL SYMG (G,R,N,NFILE)
C
      REWIND NFILE
      RETURN
C     $$$$$$$$$$$$$$$$$$$$$$$$$$$$$$$$$$$$$$$$$$$$$$$$$$$$$$$$$$$$
C     $                                                          $
C     $     CALLED SUBROUTINES   ....   LOC,UNPAC                $
C     $                                                          $
C     $$$$$$$$$$$$$$$$$$$$$$$$$$$$$$$$$$$$$$$$$$$$$$$$$$$$$$$$$$$$
      END




      SUBROUTINE SYMG(H,G,N,NFILE)
C     $$$$$$$$$$$$$$$$$$$$$$$$$$$$$$$$$$$$$$$$$$$$$$$$$$$$$$$$$$$$
C     $   THIS ROUTINE TAKES A HF MATRIX WHICH HAS BEEN          $
C     $  FORMED FROM ONLY THE 'POINT GROUP DISTINCT'             $
C     $  ONE AND TWO ELECTRON INTEGRALS AND APPLIES THE          $
C     $  OPERATIONS OF THE GROUP TO GENERATE THE FULL HF         $
C     $  MATRIX. FOR THEORY SEE P.D.DACRE CHEM. PHYS. LETT.      $
C     $   VOL. 7 P47  (1970)                                     $
C     $                                                          $
C     $$$$$$$$$$$$$$$$$$$$$$$$$$$$$$$$$$$$$$$$$$$$$$$$$$$$$$$$$$$$
C**********************************************************************C
C       ARGUMENTS                                                      C
C           H   INPUT  HF OUTPUT FULL HF       ( STORAGE MODE 1)       C
C           G   WORKSPACE MATRIX - THIS MATRIX IS 'R' WHEN SYMG IS     C
C                  CALLED FROM GOFR SO THE ROUTINE SCFCS MUST RESTORE  C
C                  THE 'R' MATRIX FROM DISK TO ENSURE COMPATIBILITY    C
C                  WITH THE USE OF GOFR WITHOUT SYMG                   C
C           N   THE DIMENSION OF H,G - BASIS SIZE                      C
C           NFILE  THE LOGICAL NUMBER OF THE TWO ELECTRON INTEGRAL     C
C                  FILE - THE SYMMETRY MATRIX MTRAN IS PICKED UP FROM  C
C                  THE END                                             C
C**********************************************************************C
      DIMENSION H(1),G(1),MTRAN(40,24)
      DATA JUMP/1/
      NS=N*(N+1)/2
C  ....  READ MTRAN ONLY ON FIRST ENTRY INTO THIS ROUTINE AND SAVE IT
      GO TO (7,8),JUMP
    7 JUMP=2
      READ(NFILE) MMAX,MTRAN
    8 DO 10 I=1,NS
   10 G(I)=H(I)
      DO 1 M=1,MMAX
      DO 1 I=1,N
      IT=IABS(MTRAN(I,M))
      SIGNI=1.0
      IF(MTRAN(I,M)) 2,2,3
    2 SIGNI=-1.0
    3 DO 1 J=1,I
      JT=IABS(MTRAN(J,M))
      SIGNJ=1.0
      IF(MTRAN(J,M)) 4,4,5
    4 SIGNJ=-1.0
    5 IJ=I*(I-1)/2+J
      CALL LOC(IT,JT,ITJT,N,N,1)
    1 H(ITJT)=H(ITJT)+SIGNI*SIGNJ*G(IJ)
      RETURN
      END
```

```
      SUBROUTINE EIGEB(A,R,N,MV,IBLOC,NBLOC)
C     ....  THIS ROUTINE IS BASED ON THE IBM SSP 'EIGEN'
C     ....  DIAGONALISE A BLOCKED MATRIX BY THE JACOBI METHOD
C     ....  THE MATRIX IS ASSUMED TO CONSIST OF A SERIES OF
C     ....  SUB-MATRICES ALONG THE MAIN DIAGONAL
C     ....   THERE ARE NBLOC SUB-MATRICES OF DIMENSION IBLOC(1) .....
C     ....   ...IBLOC(NBLOC)
C     ....   SET NBLOC=1,IBLOC(1)=N FOR NORMAL USE
C*********************************************************************** C
C    ARGUMENTS                                                          C
C           A   THE MATRIX TO BE DIAGONALISED STORAGE MODE 1            C
C                   THIS MATRIX IS DESTROYED - THE EIGENVALUES ARE      C
C                   PUT ONTO THE MAIN DIAGONAL, ZEROES ELSEWHERE        C
C           R    THE MATRIX OF EIGENVECTORS PRODUCED  STORAGE MODE 0    C
C           N     THE OVERALL DIMENSION OF THE MATRIX  A                C
C            MV    MV=0 FOR VALUES AND VECTORS MV=1 FOR VALUES ONLY     C
C           IBLOC     AS  ABOVE                                         C
C           NBLOC  AS  ABOVE                                            C
C***********************************************************************C
      DIMENSION IBLOC(1)
      DIMENSION A(1),R(1)
C     ....  CHANGE THESE CONSTANTS TO DOUBLE PRECISION WHEN CHANGING
C     ....  TO REAL*8
      DATA SMALL/1.0E-06/
      DATA ZERO,HALF,ONE,TWO,ROOT2/0.0,0.5,1.0,2.0,1.414/
C     ....  REMOVE THE FOLLOWING TWO CARDS TO CHANGE TO REAL*8
      DSQRT(XY)=SQRT(XY)
      DABS(XY)=ABS(XY)
      IF(MV-1) 10,25,10
   10 IQ=-N
C         GENERATE IDENTITY MATRIX
      DO 20 J=1,N
      IQ=IQ+N
      DO 20 I=1,N
      IJ=IQ+I
      R(IJ)=ZERO
      IF(I-J) 20,15,20
   15 R(IJ)=ONE
   20 CONTINUE
C         COMPUTE INITIAL AND FINAL NORMS (ANORM AND ANORMX)
   25 ANORM=ZERO
      DO 35 I=1,N
      DO 35 J=I,N
      IF(I-J) 30,35,30
   30 IA=I+(J*J-J)/2
      ANORM=ANORM+A(IA)*A(IA)
   35 CONTINUE
      IF(ANORM) 165,165,40
   40 ANORM=ROOT2*DSQRT(ANORM)
      ANRMX=ANORM*SMALL/FLOAT(N)
C         INITIALIZE INDICATORS AND COMPUTE THRESHOLD, THR
      IMAX=0
      DO 335 ISYM=1,NBLOC
      IMIN=IMAX+1
      IMAX=IMAX+IBLOC(ISYM)
      IND=0
      THR=ANORM
   45 THR=THR/FLOAT(N)
   50 L=IMIN
   55 M=L+1
C     ....  COMPUTE THE SIN AND COS OF THE ROTATION WHICH TRANSFORMS
C     ....  ELEMENT A(LM) TO ZERO
   60 MQ=(M*M-M)/2
      LQ=(L*L-L)/2
      LM=L+MQ
   62 IF(DABS(A(LM))-THR) 130,65,65
   65 IND=1
      LL=L+LQ
      MM=M+MQ
      X=HALF*(A(LL)-A(MM))
   68 Y=-A(LM)/DSQRT(A(LM)*A(LM)+X*X)
      IF(X) 70,75,75
   70 Y=-Y
   75 SINX=Y/DSQRT(TWO*(ONE+(DSQRT(ONE-Y*Y))))
      SINX2=SINX*SINX
```

```
   78 COSX=DSQRT(ONE-SINX2)
      COSX2=COSX*COSX
      SINCS =SINX*COSX
C         ROTATE L AND M COLUMNS
      ILQ=N*(L-1)
      IMQ=N*(M-1)
      DO 125 I=1,N
      IQ=(I*I-I)/2
      IF(I-L) 80,115,80
   80 IF(I-M) 85,115,90
   85 IM=I+MQ
      GO TO 95
   90 IM=M+IQ
   95 IF(I-L) 100,105,105
  100 IL=I+LQ
      GO TO 110
  105 IL=L+IQ
  110 X=A(IL)*COSX-A(IM)*SINX
      A(IM)=A(IL)*SINX+A(IM)*COSX
      A(IL)=X
  115 IF(MV-1) 120,125,120
  120 ILR=ILQ+I
      IMR=IMQ+I
      X=R(ILR)*COSX-R(IMR)*SINX
      R(IMR)=R(ILR)*SINX+R(IMR)*COSX
      R(ILR)=X
  125 CONTINUE
      X=TWO*A(LM)*SINCS
      Y=A(LL)*COSX2+A(MM)*SINX2-X
      X=A(LL)*SINX2+A(MM)*COSX2+X
      A(LM)=(A(LL)-A(MM))*SINCS+A(LM)*(COSX2-SINX2)
      A(LL)=Y
      A(MM)=X
C         TESTS FOR COMPLETION
C         TEST FOR M = LAST COLUMN
  130 IF(M-IMAX) 135,140,135
  135 M=M+1
      GO TO 60
C         TEST FOR L = SECOND FROM LAST COLUMN
  140 IF(L-(IMAX-1)) 145,150,145
  145 L=L+1
      GO TO 55
  150 IF(IND-1) 160,155,160
  155 IND=0
      GO TO 50
C         COMPARE THRESHOLD WITH FINAL NORM
  160 IF(THR-ANRMX) 335,335,45
  335 CONTINUE
C         SORT EIGENVALUES AND EIGENVECTORS
  165 IQ=-N
      DO 185 I=1,N
      IQ=IQ+N
      LL=I+(I*I-I)/2
      JQ=N*(I-2)
      DO 185 J=I,N
      JQ=JQ+N
      MM=J+(J*J-J)/2
      IF(A(LL)-A(MM)) 185,185,170
  170 X=A(LL)
      A(LL)=A(MM)
      A(MM)=X
      IF(MV-1) 175,185,175
  175 DO 180 K=1,N
      ILR=IQ+K
      IMR=JQ+K
      X=R(ILR)
      R(ILR)=R(IMR)
  180 R(IMR)=X
  185 CONTINUE
      RETURN
      END
```

```
      SUBROUTINE MM(B,C,COL,N)
C  ....  MATRIX MULTIPLICATION USING VECTOR WORKSPACE
C  ....       C = B * C
C  ....    B  UNCHANGED
C     STORAGE MODE OF B,C MUST BE 0
C  ....  THIS VERSION WORKS FOR SQUARE MATRICES ONLY
C***********************************************************************C
C     ARGUMENTS                                                         C
C          B  LEFT HAND FACTOR IN PRODUCT                               C
C          C  RIGHT HAND FACTOR IN PRODUCT  ALSO OUTPUT  B*C            C
C          COL  WORKSPACE, MUST BE DIMENSIONED TO N                     C
C          N  DIMENSION OF THE (SQUARE) MATRICES                        C
C***********************************************************************C
C  ....  STORAGE MODE OF B,C  MUST BE 0 SINCE PRODUCT OF TWO
C  ....  SYMMETRICAL MATRICES IS NOT NESC. SYMMETRICAL
      DIMENSION B(1),C(1),COL(1)
      DATA ZERO/0.0/
      DO 3 J=1,N
      DO 2 I=1,N
      AA=ZERO
      DO 1 K=1,N
      CALL LOC(I,K,IK,N,N,0)
      CALL LOC(K,J,KJ,N,N,0)
    1 AA=AA+B(IK)*C(KJ)
    2 COL(I)=AA
      DO 3 K=1,N
      CALL LOC(K,J,KJ,N,N,0)
    3 C(KJ)=COL(K)
      RETURN
      END


      SUBROUTINE VDGHV(H,V,COL,N)
C  ....  ONE ELECTRON TRANSFORMATION  SETS H,V = V(DAGGER)*H*V
C  ....  SAME METHOD AS IN  MM
C  ....  H,V MUST HAVE STORAGE MODE 0
C  ....  BOTH MATRICES ARE DESTROYED - SO BOTH ARE EQUAL ON OUTPUT
C***********************************************************************C
C    ARGUMENTS                                                          C
C          H  INPUT MATRIX TO BE TRANSFORMED . CONTAINS V(DAG)*H*V ON   C
C                OUTPUT                                                 C
C           V  TRANSFORMATION MATRIX CONTAINS V(DAG)*H*V ON OUTPUT      C
C          COL  WORKSPACE - DIMENSIONED TO N                            C
C          N   DIMENSION OF THE SQUARE MATRICES H V                     C
C***********************************************************************C
      DIMENSION H(1),V(1),COL(1)
      DATA ZERO/0.0/
      DO 3 J=1,N
      DO 2 I=1,N
      AA=ZERO
      DO 1 K=1,N
      CALL LOC(K,I,KI,N,N,0)
      CALL LOC(K,J,KJ,N,N,0)
    1 AA=AA+V(KI)*H(KJ)
    2 COL(I)=AA
      DO 3 K=1,N
      CALL LOC(K,J,KJ,N,N,0)
    3 H(KJ)=COL(K)
      DO 33 J=1,N
      DO 22 I=1,N
      AA=ZERO
      DO 11 K=1,N
      CALL LOC(I,K,IK,N,N,0)
      CALL LOC(K,J,KJ,N,N,0)
   11 AA=AA+H(IK)*V(KJ)
   22 COL(I)=AA
      DO 33 K=1,N
      CALL LOC(K,J,KJ,N,N,0)
   33 V(KJ)=COL(K)
      NN=N*N
      DO 10 I=1,NN
   10 H(I)=V(I)
      RETURN
      END
```

```
      SUBROUTINE RMAT(U,R,N,MIN,MAX)
C  ....  FORMS THE 'R' MATRIX FROM THE COLUMNS MIN TO MAX OF U
C  ....  IF U IS UNITARY THEN R IS IDEMPOTENT
C**********************************************************************C
C     ARGUMENTS                                                        C
C           U   EIGENVECTOR MATRIX - STORAGE MODE 0                    C
C           R   OUTPUT DENSITY MATRIX - STORAGE MODE 1                 C
C           N   DIMENSION OF U,R                                       C
C           MIN   FIRST COLUMN OF U USED TO FORM R                     C
C           MAX   LAST COLUMN OF U IN DEFINITION OF R                  C
C**********************************************************************C
      DIMENSION U(1),R(1)
      DATA ZERO/0.0/
      DO 1 I=1,N
      DO 1 K=1,I
      AA=ZERO
      IK=I*(I-1)/2+K
      DO 2 J=MIN,MAX
      CALL LOC(I,J,IJ,N,N,0)
      CALL LOC(K,J,KJ,N,N,0)
    2 AA=AA+U(IJ)*U(KJ)
      CALL LOC(I,K,IK,N,N,1)
    1 R(IK)=AA
      RETURN
      END
```

```
      SUBROUTINE TRACE(A,B,N,SUM)
C  ....  ROUTINE TO ADD (REPEAT ADD) TRACE(A*B) INTO SUM
C  ....  USAGE IS FORMATION OF THE ELECTRONIC ENERGY IN HF CALCS
C**********************************************************************C
C        ARGUMENTS                                                     C
C           A   ONE FACTOR IN PRODUCT MATRIX                           C
C           B   OTHER FACTOR IN PRODUCT                                C
C                  A,B  STORAGE MODE 1                                 C
C           N   DIMENSION OF A,B                                       C
C           SUM   ACCUMULATOR FOR TRACE                                C
C**********************************************************************C
      DIMENSION A(1),B(1)
      DO 1 I=1,N
      DO 1 J=1,I
      IJ=I*(I-1)/2+J
      TERM=A(IJ)*B(IJ)
      IF(I-J) 2,1,2
    2 TERM=TERM+TERM
    1 SUM=SUM+TERM
      RETURN
      END
```

```
      SUBROUTINE VFORM (A,B,N,IBLOC,NBLOC,NSCRA,ISCRA,IREAD,IRITE)
C  ....   ROUTINE TO ASSEMBLE THE SYMMETRY ORBITAL MATRIX AND
C  ....  TO COMPUTE THE MATRIX  ....
C  ....        V = T*(TDAG*S*T)**0.5
C  ....  WHICH ORTHONORMALISES AND SYMMETRY BLOCKS THE HF MATRIX
C  ....  WHERE T FORMS SYMMETRY ORBITALS AND  S  IS THE 'AO' OVERLAP
C  ....  MATRIX -  CONVENTION THROUGHOUT IS  ........
C  ....  BASIS FUNCTIONS FORM A ROW MATRIX
C  ....   LINEAR COEFFICIENTS FORM A COLUMN MATRIX
C  ....  THIS ENSURES THAT THE MATRIX SCHROEDINGER EQUATION IS
C  ....          HF*C = E*C      AS      EXPECTED
      DIMENSION A(600),B(600),IBLOC(1),COL(50)
C  ....  A SENSIBLE MODIFICATION WOULD BE TO MAKE THE MATRICES A,B OF
C  ....  'VARIABLE' DIMENSION -ADDING AN ARGUMENT TO THE LIST
C  ....       IN THIS  CASE N*N
C**********************************************************************C
C        ARGUMENTS                                                     C
C          A     THE OVERLAP MATRIX - STORAGE MODE  1                  C
C          B       OUTPUT   V  MATRIX  - STORAGE MODE  0               C
C          N   THE DIMENSION OF A,B                                    C
C          IBLOC     A VECTOR TO CONTAIN THE 'BLOCKING INFORMATION'    C
C                  READ IN  - THE MATRIX  ASSEMBLED INTO  B MUST       C
C                BLOCK THE S MATRIX INTO SUB-MATRICES OF DIMENSION     C
C                IBLOC(1),  ..  IBLOC(NBLOC)                           C
C          NBLOC  THE NUMBER OF SUB-MATRICES  SEE ABOVE                C
C          NSCRA  THE LOGICAL NUMBER OF A SCRATCH FILE                 C
C          ISCRA    ASSOCIATED VARIABLE FOR NSCRA                      C
C          IREAD  THE NUMBER OF A READER - ZERO IF THE T MATRIX IS     C
C                THE UNIT MATRIX                                       C
C  ....  THE DATA TO FORM THE T MATRIX IS IN THE FORM .........        C
C  ....  NE,VAL,CON   (I3,2F10.6)                                      C
C  ....   (I(K),J(K),K=1,NE)   (26I3)                                  C
C  ....  WHERE THE I,J ELEMENTS OF T ARE  T(I J)=CON/SQRT(VAL)         C
C  ....  THIS IS REPEATED UNTIL ALL NON-ZERO ELEMENTS HAVE BEEN        C
C  ....  ASSEMBLED  -  DATA IS TERMINATED BY A BLANK CARD              C
C          IRITE    THE NUMBER OF A PRINTER - ZERO IF THE T MATRIX     C
C               IS NOT TO BE PRINTED                                   C
C**********************************************************************C
      DIMENSION K(100)
      DATA  T/'T'/
C  ....  CONSTANTS ARE PUT INTO 'DATA' STATEMENTS TO FACILITATE
C  ....  CHANGEOVER TO DOUBLE PRECISION  (IMPLICIT REAL*8)
      DATA ZERO,ONE/0.0,1.0/
C  ....  REMOVE THE FOLLOWING CARD WHEN CHANGING TO REAL*8
      DSQRT(X)=SQRT(X)
      WRITE(NSCRA'1) A
      ITPOS=ISCRA
      NN=N*N
      DO 5 I=1,NN
    5 B(I)=ZERO
      IF(IREAD) 7,7,6
C  ....  IF IREAD IS ZERO SET T=1
    7 DO 8 I=1,N
      CALL LOC(I,I,II,N,N,0)
    8 B(II)=ONE
      NBLOC=1
      IBLOC(1)=N
      GO TO 1
C  ....  OTHERWISE READ IN THE NON ZERO ELEMENTS OF   T
    6 CONTINUE
      READ(IREAD,100) NBLOC,(IBLOC(I),I=1,NBLOC)
  100 FORMAT(26I3)
    3 READ(IREAD,101) NE,VAL,CON
      VAL=CON/DSQRT(VAL)
  101 FORMAT(I3,2F10.5)
      IF(NE)1,1,2
    2 NE2=NE+NE
      READ(IREAD,100) (K(I),I=1,NE2)
      DO 4 I=2,NE2,2
      ID=K(I-1)
      JD=K(I)
      CALL LOC(ID,JD,IJ,N,N,0)
    4 B(IJ)=VAL
      GO TO 3
    1 WRITE(NSCRA'ITPOS)B
```

```
      IF(IRITE) 11,11,10
   10 CALL MATP(B,N,N,N,T,IRITE)
   11 CONTINUE
      FIND(NSCRA'ITPOS)
      CALL MODE(A,N,1,0)
C  ....  FORM TDAG*S*T   -IT SHOULD BE BLOCKED ACCORDING TO IBLOC
      CALL VDGHV(A,B,COL,N)
      CALL MODE(B,N,0,1)
C  ....  FORM (TDAG*S*T)**-0.5     USING SHALF
      CALL SHALF(B,A,COL,N,IBLOC,NBLOC)
      READ(NSCRA'ITPOS) A
C  ....  FORM THE FINAL TRANSFORMATION MATRIX  ....
C  ....       V = T*(TDAG*S*T)**-0.5
      CALL MM(A,B,COL,N)
C  ....   RESTORE THE ARGUMENT  A
      READ(NSCRA'1)A
      RETURN
      END



      SUBROUTINE SHALF(A,U,COL,N,IBLOC,NBLOC)
C  ....  CALCULATION OF A**(-0.5)   FOR ORTHOGONALISATION
C  ....        NO CHECKS FOR ZERO EIGENVALUES  SO LOOK  OUT
C*************************************************************************C
C          ARGUMENTS                                                      C
C          A    A  IS  REPLACED BY A**-0.5                                C
C                  INPUT STORAGE MODE 1, OUTPUT STORAGE MODE 0            C
C           U  IS  WORK SPACE  FOR THE DIAGONALISATION                    C
C            N   DIMENSION OF A,U                                         C
C            IBLOC  A VECTOR CONTAINING THE BLOCKING OF MATRIX A          C
C               NBLOC    IT IS ASSUMED THAT THERE ARE NBLOC SUB MATRICES  C
C                    OF DIMENSION IBLOC(1) . . . IBLOC(NBLOC) ALONG THE   C
C                    DIAGONAL OF A   SET IBLOC(1)=N,NBLOC=N FOR NORMAL USE C
C*************************************************************************C
      DIMENSION A(1),U(1),IBLOC(1)
      DIMENSION COL(1)
      DATA ONE/1.0/
C  ....  REMOVE THE FOLLOWING CARD WHEN CHANGING TO IMPLICIT REAL*8
      DSQRT(X)=SQRT(X)
      CALL EIGEB(A,U,N,0,IBLOC,NBLOC)
      CALL STRA(U,N)
      DO 1 I=1,N
      II=I*(I+1)/2
    1 A(II)=ONE/DSQRT(A(II))
      CALL MODE(A,N,1,0)
      CALL VDGHV(A,U,COL,N)
      RETURN
      END
```

```
      SUBROUTINE MATP(A,M,N,IA,TITLE,JP)
C     ....  MATRIX PRINTER  ADAPTED FROM ATLAS LIBRARY
C     ....  WRITES OUT THE MATRIX  A  ON  UNIT  JP  A  MAY BE SINGLY
C     ....  OR DOUBLY SUBSCRIPTED / IF A IS SINGLY SUBSCRIPTED IT MUST BE
C     ....  STORAGE  MODE  0
C*********************************************************************C
C     ARGUMENTS                                                       C
C           A   MATRIX TO BE PRINTED                                  C
C           M,N   NUMBER OF ROWS  AND  COLUMNS IN  A                  C
C           IA  THIS INTEGER VARIABLE MUST BE SET TO  M  IF THE       C
C               MATRIX IS SINGLE SUBSCRIPTED (MODE 0) OR TO THE       C
C            NUMBER OF ROWS OF A IN THE DIMENSION STATEMENT OF THE    C
C                 CALLING PROGRAM IF A IS DOUBLY SUBSCRIPTED          C
C           TITLE     THIS VARIABLE SHOULD CONTAIN A TITLE IN A4 FORMAT C
C           JP   THE OUTPUT DEVICE (PRINTER) LOGICAL UNIT NUMBER      C
C*********************************************************************C
      DIMENSION A(1)
      IF(JP) 99,99,20
   20 CONTINUE
      IP=1
      J3=(N-1)/5+1
      DO 101 J=1,J3
      J1=5*J-4
      J2=5*J
      IF(J2-N)6,6,7
    7 J2=N
   14 FORMAT (1H1,//,10X,A4,' MATRIX  ',4X,'PART  ',I3,////)
    6 WRITE(JP,14) TITLE,IP
      IP=IP+1
      WRITE(JP,4) (J4,J4=J1,J2)
    4 FORMAT (I19,4I20)
      WRITE (JP,3)
    3 FORMAT (1X)
      M1=1
      M2=1
      M3=0
      DO 1 I=1,M
      K1=IA*(J1-1)+I
      K2=IA*(J2-1)+I
      WRITE (JP,5) I,(A(K),K=K1,K2,IA)
      WRITE(JP,3)
    5 FORMAT (I5,5F20.7)
      IF (M1-5)9,10,9
10    WRITE (JP,8)
    8 FORMAT (1X)
      M1=0
    9 IF (M2-30)11,12,11
   12 IF (M2-M) 13,1,1
   13 WRITE(JP,14) TITLE,IP
      WRIT E(        JP,4)   (J4,J4=J1,J2)
      IP=IP+1
      M2=0
   11 M1=M1+1
      M2=M2+1
    1 CONTINUE
  101 CONTINUE
   99 CONTINUE
      RETURN
      END
```

```
      SUBROUTINE STRA(A,N)
C   ....   TRANSPOSITION OF A MATRIX 'A' IN ITS OWN SPACE
C   ....  A IS STORAGE MODE 0 OF COURSE
C   ....   FOR USE IN CONJUNCTION WITH E.G. VDGHV
C**********************************************************************C
C       ARGUMENTS                                                      C
C             A    A   IS REPLACED BY A(DAGGER)                        C
C            N   DIMENSION OF  A                                       C
C**********************************************************************C
      DIMENSION A(1)
      DO 1 I=1,N
      DO 1 J=1,I
      CALL LOC(I,J,IJ,N,N,0)
      CALL LOC(J,I,JI,N,N,0)
      D=A(IJ)
      A(IJ)=A(JI)
    1 A(JI)=D
      RETURN
      END
```

```
PACK     START 0
* PACK IS THE ROUTINE CALLED BY  ''CALL PACK(A,I,J,K,L)''
*  IT PUTS THE LEAST SIGNIFICANT BYTE OF I,J,K,L INTO A THUS STORING
*     FOUR INTEGERS (VALUES UP TO 255) IN THE SPACE OF ONE REAL
*        VARIABLE
*     WRITTEN (UNDER DURESS) BY M. ELDER
* THE ENTRY ADRESS IS IN REGISTER 15
* THE RETURN ADDRESS IS IN REGISTER 14
* THE ADDRESS OF THE ARGUMENT LIST IS IN REGISTER 1
* THE ADDRESS OF THE SAVE AREA IS IN REGISTER 13
         BC    15,12(15)                BRANCH AROUND NEXT TWO LINES
         DC    X'7'
         DC    CL7'PACK
         STM   14,3,12(13)              SAVE REGISTERS 14,15,0,1,2,3
         L     3,16(1)                  PUT THE ADDRESS OF L IN 3
         IC    2,3(3)                   PUT THE LOW ORDER BYTE OF L IN 2
         SLL   2,8(0)                   SHIFT LEFT 8 BITS READY FOR K
         L     3,12(1)                  REPEAT FOR K,J AND I
         IC    2,3(3)
         SLL   2,8(0)
         L     3,8(1)
         IC    2,3(3)
         SLL   2,8(0)
         L     3,4(1)
         IC    2,3(3)
         L     3,0(1)                   PUT THE ADDRESS OF A IN 3
         ST    2,0(3)                   STORE THE PACKED WORD IN 2 AT A
         LM    2,3,28(13)               RESTORE REGISTERS
         MVI   12(13),X'FF'             SET BYTE 13 OF SAVE AREA TO 1'S
         BCR   15,14                    RETURN
         END

UNPAC    START 0
* UNPAC  IS THE ROUTINE CALLED BY ''CALL UNPAC (A,I,J,K,L)''
*  THIS ROUTINE TAKES THE FOUR BYTES IN  A  AND PUTS THEM INTO
*     THE LOW ORDER POSITIONS OF I,J,K,L - REVERSING THE ACTION
*       OF 'PACK'
* SEE PACK LISTING FOR DETAILS OF LINKAGE CONVENTIONS
         BC    15,12(15)
         DC    X'7'
         DC    CL7'UNPAC
         STM   14,3,12(13)
         L     3,0(1)                   PUT THE ADDRESS OF A INTO 3
         L     2,0(3)                   LOAD A INTO 2
         L     3,4(1)                   LOAD THE ADDRESS OF I INTO 3
         STC   2,3(3)                   STORE THE LOW ORDER BYTE OF A IN
*                                        THE LOW ORDER BYTE OF I
         SRL   2,8(0)                   SHIFT RIGHT 8 BITS READY FOR J
         L     3,8(1)                   REPEAT FOR J,K AND L
         STC   2,3(3)
         SRL   2,8(0)
         L     3,12(1)
         STC   2,3(3)
         SRL   2,8(0)
         L     3,16(1)
         STC   2,3(3)
         LM    2,3,28(13)
         MVI   12(13),X'FF'
         BCR   15,14
         END
```

```
      SUBROUTINE GOSFR(R,G,N,NFILE,IFILE,EM1,EN1,EM2,EN2)
C  ....  FORMATION OF 'G' MATRICES FOR OPEN SHELL HARTREE-FOCK
      REAL LABEL(200)
      DIMENSION VALUE(200)
C***********************************************************************
C                                                                     C
C          ARGUMENTS                                                  C
C             R    A VECTOR DIMENSIONED TO AT LEAST N*(N+1)  - THE    C
C                  FIRST N*(N+1)/2 ELEMENTS ARE THE FIRST  R MATRIX   C
C                        STORAGE MODE 1 OF COURSE                     C
C                    THE NEXT N*(N+1)/2 ELEMENTS ARE THE OTHER R  MATRIX C
C                   STORAGE MODE 1                                    C
C             G      THE RESULTANT  G(R) MATRICES ARE ADDED TO THE    C
C                     TWO STORAGE MODE 1 MATRICES IN THE N*(N+1) ELEMENTS C
C                      OF  G                                          C
C                                                                     C
C             N        BASIS SIZE                                     C
C             NFILE    LOGICAL NUMBER OF THE TWO ELECTRON INTEGRAL FILE C
C             IFILE  REDUNDANT  - ASSOCIATED VARIABLE FOR NFILE IN DISK C
C                        APPLICATIONS                                 C
C             EM1,EN1,EM2,EN2        .......                          C
C     $$$$$$$$$$$$$$$$$$$$$$$$$$$$$$$$$$$$$$$$$$$$$$$$$$$$$$$$$$$$
C     $     THE MATRICES FORMED ARE    ...                       $
C     $                                                          $
C     $    G1(R1) = EM1*J(R1) - EN1*K(R1)                        $
C     $                                                          $
C     $  AND    G2(R2) = EM2*J(R2) - EN2*K(R2)                   $
C     $                                                          $
C     $    WHERE THE TEMPORARY NOTATION ' 1 & 2 ' HAS BEEN       $
C     $    USED TO DENOTE THE FIRST AND SECOND MATRICES          $
C     $      STORED IN  R  AND  G  RESPECTIVELY                  $
C     $$$$$$$$$$$$$$$$$$$$$$$$$$$$$$$$$$$$$$$$$$$$$$$$$$$$$$$$$$$$
C                                                                     C
C***********************************************************************
      DIMENSION R(1),G(1)
      NS=N*(N+1)/2
       IFILE =1
C  ....  FOR DISK APPLICATIONS TAKE THE 'C' FROM THE FOLLOWING CARD
C  51 READ(NFILE'IFILE) NN,IEND,LABEL,VALUE
C  ....  AND THROW AWAY THE PRESENT STATEMENT 51
   51 READ(NFILE) NN,IEND,LABEL,VALUE
      DO 29 M=1,NN
      CALL UNPAC(LABEL(M),I,J,K,L)
      VA=VALUE(M)
      VAL =VA*EM1
      VAL2 =VA*EM2
      RN=VA*EN1
      RN2=VA*EN2
C  ....  'REVERSE DICTIONARY' ORDER
C  ....  COMPUTE THE SUBSCRIPTS FOR THE CONTRIBUTIONS OF I,J,K,L
C  ....   TO THE  J AND  K  MATRICES
      II=I*(I-1)/2
      KK=K*(K-1)/2
      IJ=II+J
      KL=KK+L
      IL=II+L
      IK=II+K
      CALL LOC(J,K,JK,N,N,1)
      CALL LOC(J,L,JL,N,N,1)
      IJ2=IJ+NS
      KL2=KL+NS
      JK2=JK+NS
      IK2=IK+NS
      IL2=IL+NS
      JL2=JL+NS
      DA=R(IJ)*VAL
      DA2=R(IJ2)*VAL2
      DB=R(KL)*VAL
      DB2=R(KL2)*VAL2
      SV=R(JK)*RN
      SV2=R(JK2)*RN2
      SA=R(IK)*RN
      SA2=R(IK2)*RN2
      SB=R(IL)*RN
      SB2=R(IL2)*RN2
      SC=R(JL)*RN
```

```
      SC2=R(JL2)*RN2
      IF(K-L) 99,97,99
   99 DB=DB+DB
      DB2=DB2+DB2
      G(IK)=G(IK)-SC
      G(IK2)=G(IK2)-SC2
      IF(I-J) 96,97,96
   96 IF(J-K) 97,95,95
   95 G(JK)=G(JK) -SB
      G(JK2)=G(JK2)-SB2
   97 G(IL)=G(IL)-SV
      G(IL2)=G(IL2)-SV2
      G(IJ)=G(IJ)+DB
      G(IJ2)=G(IJ2)+DB2
      IF(I-J) 14,15,14
   14 IF(J-L) 15,12,12
   12 G(JL)=G(JL)-SA
      G(JL2)=G(JL2)-SA2
   15 IF(IJ-KL) 21,29,21
   21 CON=DA
      CON2=DA2
      IF(I-J) 56,57,56
   56 CON=CON+DA
      CON2=CON2+DA2
   57 IF(J-K) 61,61,22
   61 G(JK)=G(JK)-SB
      G(JK2)=G(JK2)-SB2
      IF(I-J) 59,62,59
   59 IF(I-K) 60,60,62
   60 G(IK)=G(IK)-SC
      G(IK2)=G(IK2)-SC2
   62 IF(K-L) 74,22,74
   74 IF(J-L) 73,73,22
   73 G(JL)=G(JL)-SA
      G(JL2)=G(JL2)-SA2
   22 G(KL)=G(KL)+CON
      G(KL2)=G(KL2)+CON2
   29 CONTINUE
      IF(IEND) 53,51,53
   53 CONTINUE
      REWIND NFILE
      RETURN
C
C     $$$$$$$$$$$$$$$$$$$$$$$$$$$$$$$$$$$$$$$$$$$$$$$$$$$$$$$$$$$$
C     $                                                          $
C     $     CALLED SUBROUTINES  ....  LOC,UNPAC                  $
C     $                                                          $
C     $$$$$$$$$$$$$$$$$$$$$$$$$$$$$$$$$$$$$$$$$$$$$$$$$$$$$$$$$$$$
C
      END
```

INDEX

The subject matter of this volume is too restricted to merit a detailed subject index. However, many mathematical symbols have been used throughout the text and many of these have been used with constant meaning. These symbols are indexed here.

α, A (alpha)
β, B (beta)
Γ, γ (gamma)
δ, Δ (delta)
ε, E (epsilon)
ζ, Z (zeta)
η, H (eta)
θ, Θ (theta)
ι, I (iota)
κ, K (kappa)
λ, Λ (lambda)
μ, M (mu)
ν, N (nu)
ξ, Ξ (xi)
ο, O (omicron)
π, Π (pi)
ρ, P (rho)
σ, Σ (sigma)
τ, T (tau)
υ, T (upsilon)
ϕ, Φ (phi)
χ, X (chi)
ψ, Ψ (psi)
ω, Ω (omega)

The Greek alphabet is given above for reference purposes.

INDEX OF MATRICES

All matrices are written in large letters - A, B - and their elements in standard text with the corresponding subscripts - A_{ij}, B_{21}.

Symbol	Main use	Definition in section
A	Definition of optimum hybrids	13.4
B	A, B are used as arbitrary matrices	
$c^{(i)}$	The coefficients of the ith AO - the ith column of C	4.2
C	The transformation from basis functions to AOs	4.2
D	Column of linear coefficients in VB theory	3.7
$\mathsf{D}^{(i)}$	Matrix representation of a symmetry operation	12.1
E	The matrix which diagonalises S	9.7
G, $\mathsf{G}(\mathsf{R})$	Roothaan electron interaction matrix	4.2, 5.2, 11.2, 11.3
H	One-electron Hamiltonian	4.2, 5.2
H^{F}	Roothaan-Hartree-Fock (RHF) matrix	4.2, 5.2
$\mathsf{H}^{F\alpha}$, $\mathsf{H}^{F\beta}$	DODS RHF matrices	11.2
H^{FC}, H^{FO}	Open-shell RHF matrices	11.3

$\mathbf{H}$	VB Hamiltonian matrix	3.7
$\mathbf{K}$	MO "exchange" integrals	13.2
$\mathbf{L}$	Definition of localised MO's	9.3, 13.2
$\mathbf{M}$	Collection of symmetry-group information	12.2
$\mathbf{P}$	"Charge and Bond order" matrix ($2\mathbf{R}$)	10.2
$Q^{(\alpha)}$	Coefficients defining a symmetry orbital - a column of	12.5
$\mathbf{Q}$	The transformation from AO's to symmetry orbitals	12.5
$\mathbf{R}$	Charge and bond order matrix	5.2
$\mathbf{R}^{\alpha}$, $\mathbf{R}^{\beta}$	DODS $\mathbf{R}$ matrices	11.2
$\mathbf{R}_c$, $\mathbf{R}_o$	Open-shell $\mathbf{R}$ matrices	11.3
$\mathbf{S}$	Overlap matrix	4.2, 5.2
$\mathbf{T}$	Transformation from AO's to MO's	5.2
$\mathbf{T}^{\alpha}$, $\mathbf{T}^{\beta}$	DODS MO coefficients	11.2
$\mathbf{T}_o$, $\mathbf{T}_c$	Open-shell MO coefficients	11.3
$\mathbf{U}$	The matrix which diagonalises H^F	7.4
$\mathbf{U}^{\alpha}$, $\mathbf{U}^{\beta}$	DODS $\mathbf{U}$ matrix	11.2
$\mathbf{V}$	Hybridisation coefficients	9.8
$\mathbf{W}$	Core-valence orthogonalisation matrix	9.7
$\mathbf{X}$	Any orthogonalisation matrix	9.4, 9.7
$\mathbf{Y}$	Any non-singular orbital transformation	9.2
$\boldsymbol{\epsilon}$	Diagonal matrix of orbital energies	4.2, 5.2
$\boldsymbol{\chi}$	Row matrix of AO functions	4.2

$\boldsymbol{\varphi}$	Row matrix of basis functions	4.2
$\boldsymbol{\psi}$	Row matrix of MO functions	5.2

INDEX OF ROMAN SYMBOLS

Scalar quantities are written in standard text - a, r_{ij} - vector quantities in a sans serif typeface - r, r_{ij}.

Symbol	Main use	Definition in section
a,b,c	VB/SCGF linear coefficients	5.3, 13.5
$\hat{A}$	anti-symmetrising operator	3.3
A_n, B_n	"A" and "B" functions	8.3
$f(\mathsf{r})$	arbitrary function of one vector argument	3.8
$F_2(\mathsf{r}_1,\mathsf{r}_2)$,$F_n$	arbitrary functions of 2, n vector arguments	3.8
$\hat{g}(i,j)$	inter-electronic repulsion operator	2.2
$\hat{G}_i$	symmetry operation	12.1
$\hat{h}(i)$	one electron Hamiltonian	2.2
$\hat{H}$	many electron Hamiltonian	2.2
i,j,k,ℓ	subscripts to label electrons and orbitals	2.2
[i,j]	i/2(i-1) + j	8.6

(ij,kℓ) or (ij,rs)	electron repulsion integral	5.2, 8.6
k	length of GTF expansion of STO	6.3
ℓ,m,n	GTF Cartesian exponents	6.6
m	number of basis functions	4.2
n	number of electrons	2.2
n_α, n_β	number of electrons of each spin in DODS	11.2
n_c, n_o	number of electrons in Open shell MO method	11.3
P_α P_β	atomic and inter-atomic populations	10.2
P_{ij}	re-normalised changes and bond orders	10.2
P(r)	Charge density function	2.6
$\hat{P}$	Pauli permutation operator	2.3
(r,θ,ϕ)	spherical polar co-ordinates	4.2, 8.3
$r_i, r_i (=\|r_i\|)$	position vector of ith electron	2.1
$r_{\alpha i}, r_{\alpha i}$	position vector of ith electron relative to αth nucleus	2.2
r_{ij}, r_{ij}	position vector of jth electron relative to ith electron	2.2
R_α	position vector of αth nucleus	2.2
r,s,t,u	summation indices over orbitals	
s_i	"spin" variable of ith electron	2.1
$\hat{s}_i$	spin angular momentum operator of ith electron	3.3
$\hat{S}$	total electronic spin operator	3.3
$	number of unique repulsion integrals	6.2
x_i	space-spin variables for ith electron	2.1

(x_i, y_i, z_i)	Cartesian co-ordinates of ith electron	2.1
$(X_\alpha, Y_\alpha, Z_\alpha)$	Cartesian co-ordinates of αth nucleus	2.2
Z_α	charge on αth nucleus	2.2

INDEX OF GREEK SYMBOLS

Symbol*	Main use	Definition in section
$\alpha,\beta,\ldots$	subscripts to identify nuclei	2.2
$\alpha(s),\beta(s)$	spin functions	3.3
$\alpha,\beta,$	symmetry species	12.5
α	GTF exponent	6.3
$\gamma^{(\alpha)}$	symmetry orbital	12.6
δ	variation	4.2
δ_{ij}	Kronecker delta	3.5
ε_i	orbital energy	4.2, 5.2
ζ	STO orbital exponent	4.2
η	part of (ξ,η,ϕ) (prolate spheroidal) co-ordinate system	8.3
θ	part of (r,θ,ϕ) (spherical polar) co-ordinate system	4.2, 8.3
λ_i (κ_i)	spin-orbital	3.3

* The multiple use of α,β is unfortunate but in line with current use.

μ_i	spatial orbital	3.2
ξ	part of (ξ,η,ϕ) co-ordinate system	8.3
ρ_1, ρ_2	density functions	2.6
$\rho_{ij}^{(\alpha)}$	symmetry projection operator	12.5
$\sigma_{ii'}$	sign of orbital permutation	12.3
ϕ	part of (r,θ,ϕ) and (ξ,η,ϕ) coordinate systems	8.3
ϕ_i	atomic orbital	4.1
Φ	determinant of spin-orbitals	3.3
χ_i	basis function	4.2
ψ_i	molecular orbital	5.2
Ψ	n electron wave function	2.1